THE
NATURE OF
STATISTICS

W. Allen Wallis
and
Harry V. Roberts

Foreword to the Dover Edition by
George P. Shultz

Dover Publications, Inc.
Mineola, New York

Bibliographical Note

This Dover edition, first published in 2014, is an unabridged republication of the work originally published by The Free Press, New York, in 1962. This text is a revised and slightly abridged version of material forming the first quarter of *Statistics: A New Approach,* published by The Free Press, London, in 1956. A new Foreword to the Dover edition, written by George P. Shultz, has been specially prepared for the present volume.

Library of Congress Cataloging-in-Publication Data

Wallis, W. Allen (Wilson Allen), 1912–1998.
    The nature of statistics / W. Allen Wallis and Harry V. Roberts : foreword to the Dover edition by George P. Shultz. — Dover edition.
        p. cm.
    Originally published: New York : The Free Press, 1962.
    Includes bibliographical references and index.
    ISBN-13: 978-0-486-77969-0 (alk. paper)
    ISBN-10: 0-486-77969-6 (alk. paper)
     1. Statistics. I. Roberts, Harry V. II. Title.

HA29.W3356 2014
519.5—dc23

2013049278

Manufactured in the United States by Courier Corporation
77969601   2014
www.doverpublications.com

Foreword to the Dover Edition

Allen Wallis and Harry Roberts were two men of outstanding ability, common sense, and extraordinary technical competence. This book reflects these special characteristics.

Mastery, or even reasonable understanding, of statistical techniques is an important attribute to understanding what goes on in today's economy and today's world. Wallis and Roberts introduce us to these techniques, and, while this book will hardly make experts of readers, it will put them in a good position to use the techniques effectively.

The authors then draw on their extraordinary technical competence to stimulate us to take a creative approach to statistics and to offer us new insights on how to employ statistics in an illuminating and educational way. They use examples from their own experience and other events to show how to use statistics in practical, real-life situations.

Experience may be the best teacher, but only if we have learned how to observe carefully, how to organize in an effective way what we have observed, and how to bring those observations together. Wallis and Roberts invite us to conduct this exercise with them so that we are ready to conduct it on our own. As the interconnected events in our world increasingly impinge on our consciousness and on the reality that surrounds us, the need for such an exercise becomes ever more critical.

I welcome the reprinting of this classic volume, which will encourage readers to enhance their ability to observe carefully and relate their observations to one another, thereby achieving greater understanding of the reality around us.

June 2013

GEORGE P. SHULTZ

Preface

Statistics is a lively, fascinating subject, but reading about it is too often deadly dull. In this book we have tried to demonstrate the liveliness of statistics by lavish use of real examples from a wide variety of fields. We have tried to bring out its fascination by emphasizing common sense and logic and avoiding technical detail.

This is neither a how-to-do-it nor an all-about-statistics book. On the contrary, it aims to show How to Live with Statistics, Without Actually Figuring.

The Nature of Statistics is essentially a revision of the first quarter of our *Statistics: A New Approach*, published by The Free Press of Glencoe in 1956. As in that book, all tables, charts, and examples have been numbered to correspond with the pages on which they appear, a feature that will be, we hope, of much help to the reader.

The many people who helped with *Statistics: A New Approach* are thanked in its preface. It must suffice here to acknowledge again our great debts to Professors Leonard J. Savage of the University of Michigan, Frederick Mosteller of Harvard University, and William H. Kruskal of the University of Chicago.

W. Allen Wallis
Harry V. Roberts

The University of Chicago
21 March 1962

Contents

THE
NATURE OF
STATISTICS

Chapter 1

The Field of Statistics

What Is Statistics?

STATISTICS IS a body of methods for making wise decisions in the face of uncertainty.

This modern conception of the subject is a far cry from that usually held by laymen. Indeed, even the pioneers in statistical research have adopted it only within the past decade or so.

To the layman, the term "statistics" usually carries only the nebulous—and, too often, distasteful—connotation of "figures." He may even be vague about the distinction between mathematics, accounting, and statistics. In this sense, statistics are numerical descriptions of the quantitative aspects of things, and they take the form of counts or measurements. Statistics on the membership of a certain club might, for example, include a count of the number of members, and separate counts of the numbers of members of various kinds, as male and female, or over and under 21 years of age. They might include such measurements as the weights and heights of the members, or the lengths of time they can hold their breaths. Further, they might include numbers computed from such counts or measurements as those already mentioned, for example, the proportion of members who are married, the average height, or the ratios between weights and heights (that is, pounds of weight per inch of height). In this sense, the *Statistical Abstract of the United States* is a typical—and excellent—collection of statistics.

But in addition to meaning numerical facts, "statistics" refers to a subject, just as "mathematics" refers to a subject as well as to symbols, formulas, and theorems, and "accounting" refers to principles and methods as well as to accounts, balance sheets, and income statements. The subject, in this sense of statistics, is a body of methods of obtaining and analyzing data in order to base decisions on them. It is a branch of scientific method, used in dealing with phenomena that can be described numerically, either by counts or by measurements. It is in this sense that the word "statistics" is used in this book, except in the few places where the context makes it quite clear that the facts-and-figures sense is intended, for example, in the phrase "statistical data."

The purposes for which statistical data are collected can be grouped into two broad categories, which may be described loosely as practical action and scientific knowledge. Practical action here includes not only such actions by administrators as setting a bus schedule or admitting a student to school, but also such acts by individuals as having the oil changed in a car or carrying an umbrella. Scientific knowledge here includes not only knowledge gained by scientists through research, such as experiments with serums to relieve colds or analyses of records of business cycles, but also conclusions by an individual on such questions as whether coffee keeps him awake or whether his colds recur at regular intervals.

These two purposes, practical action and scientific knowledge, are by no means sharply distinct, since knowledge becomes the basis of action. For statistics, the important difference between the two purposes is that in practical action the alternatives being considered can be listed and, in principle at least, the consequences of taking each can be evaluated for each possible set of subsequent developments; whereas scientific knowledge may be employed by persons unknown for decisions not anticipated by the scientist. Thus, the consequences of error—obviously an important consideration in reaching a decision—can be taken into account more explicitly in the case of decisions for the specific "rifle-shot" purposes of practical action than in the case of decisions for the unspecified "shot-gun" purposes of scientific knowledge. The difference is, however, one of degree rather than of kind.

Statistical data, then, are collected to help decide questions of practical action or questions in scientific research. A decision about the allocation of military manpower or about a physical theory, for example, requires that the right kind of information be obtained. Statistics helps decide what kind of information is needed and how much. It then participates in the collection, tabulation, and interpretation of the data.

It is in developing methods for finding out what data mean that statisticians have evolved the present broad concept of their field. In most problems concerning the administration of business, governmental, or personal affairs, or in the search for scientific generalizations, complete information cannot be obtained; hence incomplete information must be used. Statistics provides rational principles and techniques that tell when and how judgments can be made on the basis of this partial information, and what partial information is most worth seeking. In short, statistics has come to be regarded, as we said in the first sentence, as a method of making wise decisions in the face of uncertainty.

Statistics and Scientific Method

Statistics is not a body of substantive knowledge, but a body of methods for obtaining knowledge. As such it should be viewed against the background of general methods of obtaining knowledge—of general scientific method, in short.

There is no such thing as *the* scientific method. That is, there are no procedures, formal or informal, which tell a scientist how to start, what to do next, or what conclusions to reach. Scientists rely on the same everyday methods of reasoning that are common to all intelligent problem solving. "The scientific method, as far as it is a method, is nothing more than doing one's damnedest with one's mind, no holds barred."[1]

It is enlightening, nevertheless, to recognize four stages which recur in intelligent problem-solving, or scientific method.

Four Stages in Scientific Inquiry

(1) *Observation*. The scientist observes what happens; he collects and studies facts relevant to his problem.

[1] P. W. Bridgman, "The Prospect for Intelligence," *Yale Review*, Vol. 34 (1945), pp. 444–461; quoted in James B. Conant, *On Understanding Science* (New Haven: Yale University Press, 1947), p. 115. Warren Weaver, in his presidential address to the American Association for the Advancement of Science, put the same point this way: ". . . the impressive methods that science has developed . . . involve only improvement—great, to be sure—of procedures of observation and analysis that the human race has always used. . . . In short, every man is to some degree a scientist." (*Science*, Vol. 122, 1955, p. 1258.)

(2) *Hypothesis*. To explain the facts observed, he formulates his "hunches" into a hypothesis, or theory, expressing the patterns he thinks he has detected in the data.

(3) *Prediction*. From the hypothesis or theory, he makes deductions. These, if the theory is satisfactory, constitute new knowledge, not known empirically, but deduced from the theory. If the theory is to be of value, it must make possible such new knowledge. These new facts are usually called "predictions," not in the sense of foretelling history, but rather of anticipating what will be seen if certain observations, not yet made, are made.

(4) *Verification*. He collects new facts to test the predictions made from the theory. With this step the cycle starts all over again. If the theory is substantiated, it is put to more severe tests by making more specific or more far-reaching predictions from it and testing them, until ultimately some deviation is found requiring modification of the theory. If the theory is contradicted, a new hypothesis consistent with the larger number of facts now available is formulated and then tested by steps (3) and (4); and so on. There is no final truth in science, for although failure to refute a hypothesis may increase confidence in it, no amount of testing can literally "prove" that it will always hold.

In actual scientific work these four stages are so intertwined that it would be hard to fit the history of any particular scientific investigation into such a rigid scheme. Sometimes the different stages are merged or blurred, and frequently they do not occur in the sequence listed. To know what facts to collect, one must already have some hypothesis about what facts are relevant to the problem, but such a hypothesis in turn presupposes some factual knowledge; and so forth. Nonetheless, the four stages help to focus discussion of scientific method.

Statistics is pertinent chiefly at the first and fourth stages, observation and verification, and to some extent at the second stage, formulating a hypothesis. The methods most important at the second stage, however, are primarily those of intuition, insight, imagination, and ingenuity. Very little can be said about them formally; perhaps they can be learned, but they cannot be taught. As someone has said, referring to an apocryphal story, many men noticed falling apples before Sir Isaac Newton, yet no interpretations of comparable interest were recorded by these earlier observers. The methods used at the third stage, prediction, are those of pure logic, utilizing sufficient knowledge of the field to provide those premises not given by the theory under test. The role of statistics at the first, second, and fourth stages deserves a little fuller consideration.

Statistics is helpful in the first stage, observation, because it suggests what can most advantageously be observed, and how the resulting observations can be interpreted. Not everything can be observed; it is necessary to be selective. The statistician vizualizes in detail the analysis that will be made of the observations, and the interpretation that might result from these observations. In connection with the interpretation he especially emphasizes the degree of confidence in the conclusion and the necessary allowance for error. Then he compares the different kinds and quantities of observations that could be made with the resources available, and recommends making those observations that will effect a good compromise between the conflicting goals of high confidence in the conclusions and small allowances for error.

At the second stage, statistics helps to classify, summarize, and present the results of observation in forms that are comprehensible and likely to be suggestive of fruitful hypotheses. The branch of statistics dealing with methods for doing this is called *descriptive statistics*, in contrast to *analytical statistics*, the branch dealing with methods of planning the observation of, analyzing, and basing decisions on, the data so summarized. Often, of course, summarization of important observations must necessarily be "impressionistic" or "literary" rather than numerical; this is true, for example, of anthropological studies of the character and values of cultures, or of art criticism. The statistical approach is limited to those aspects of things that can be described and summarized numerically. This limitation is not, however, as confining as it may at first appear. Many things that are "qualitative" or "subjective" nevertheless have a quantitative aspect; for example, an important aspect of a certain organic disease may be the number of times it occurs. Many subjective or qualitative impressions can be sharpened or corrected by statistical study of subsidiary details, as when the impression that racial discrimination is decreasing is checked against the number of occurrences of certain specific kinds of incident. Even though at the stage of deriving new hypotheses such extra-statistical considerations as knowledge of and intuition for the subject matter may predominate, skillful statistical organization of the materials still plays a significant role.

At the fourth stage of scientific method, hypotheses are considered verified to the extent that predictions deduced from them are borne out by later events. Sometimes, especially in the natural sciences, it is possible to speed up the testing of predictions by experimentation. Frequently, however, a prediction can be tested only by waiting to see whether it comes true; for example, some astronomical predictions forecast the course of events (history), and some medical predictions indicate what would happen to human beings under circumstances that can come about only through accident. Statistics is relevant in either situation, for the essential problem is to determine whether or not the new data observed are concordant with the prediction.

In checking a prediction with new numerical data, it is crucial to realize that the data and the prediction can seldom be expected to agree exactly, even if the theory is correct. Discrepancies may arise simply because of chance circumstances ("experimental error") that are not inconsistent with the theory. Furthermore, many important theories of modern science are probabilistic or *stochastic* rather than deterministic, in that they do not predict precisely how each observation will turn out, but only what proportion of the observations will in the long run turn out in each of a number of possible ways. Genetic theories, for example, do not in general specify the characteristics of each individual offspring of a given parentage, but only the proportions in which certain different kinds of offspring will appear. Such theories, furthermore, do not specify the proportions for any one set of observations, but only the "long-run" proportions or probabilities. In comparing a set of observations with theory, the question to be considered is, therefore, "Is the discrepancy reasonably attributable to chance?" If the discrepancy can reasonably be attributed to chance, the theory is not contradicted, and there is no adequate reason to seek special "causes" to explain the discrepancy. If the discrepancy cannot reasonably be attributed to chance, it is appropriate to look for causes—that is, to modify the theory.

Modern statistical reasoning has given a definite meaning to the verification of a hypothesis. A hypothesis is verified—"tested" is perhaps a better word—to the extent that the influence of chance in the evidence has been correctly interpreted. Statistical procedures have been evolved for measuring the risk of incorrect interpretation objectively, in terms of numerical probabilities; or, to put it differently, for measuring the risks of erroneous conclusions.

Concrete Examples of the Four Stages

Illustrations of the process just described are found in everyday experience as well as in scientific inquiries.

EXAMPLE 17 OVERHEATED CAR

(1) *Observation.* The driver of a car notices that the engine temperature is too high. (This observation might be made to verify a theory. For example, he might have observed something that made him suspect—formulate the theory—that his engine was overheated.)

(2) *Hypothesis.* He formulates the hypothesis that the fan belt is broken, and that the fan and water pump, which he knows to be driven by the fan belt, are not working for this reason.

(3) *Prediction.* From this hypothesis he deduces that the generator will not be working, since it is also driven by the fan belt, and that the ammeter will, therefore, show a zero or negative rate of charge.

(4) *Verification.* He observes the ammeter. If it shows no charging, this strengthens his confidence in the hypothesis that the fan belt is broken. It does not *prove,* however, that the fan belt is broken. Many other hypotheses are consistent with the observed data, for example, that the battery is fully charged and a regulator has stopped the charging, that something has put all the instruments out of order, and so forth.

EXAMPLE 18   THEFT OF FINISHED PRODUCT

(1) *Observation.* A certain business enterprise has to have a great deal of waste material hauled away. The net weights of four truckloads chosen at random ranged between 14,200 and 14,500 pounds.

(2) *Hypothesis.* The variation from truckload to truckload is random, in accordance with certain statistical principles that we ned not spell out here.

(3) *Prediction.* Practically all future truckloads will fall between 13,900 and 14,800 pounds. If this is true, it may result in a decision to dispense with regular weighings and pay a flat rate per truckload.

(1a) *Observation.* Several truckloads are found to weigh about 16,000 pounds. This contradicts the initial prediction and demands a new hypothesis.

(2a) *Hypothesis.* The unusually heavy truckloads may be related to trucks or drivers.

(1b) *Observation.* The heavy loads do coincide with a particular driver.

(2b) *Hypothesis.* The fact that one driver is consistently taking out unusually heavy loads, together with the already known facts that there have been shortages of the firm's finished product and that the finished product is substantially denser than the waste, suggests the hypothesis that the driver may be smuggling out finished product at the bottom of his load. More facts are required.

This example was carried out by a student during a statistics course. He got no farther with his investigation before the course ended, and we do not know what happened next. But even this much illustrates the point that actual problems go through fairly definite stages on the way to their solutions. It also illustrates an experience which is common and important: that a study started for one objective (in this example, to eliminate a work operation) may contribute to unforeseen objectives (in this case, detection of theft). Serendipity—the knack of spotting and exploiting good things encountered accidentally while searching for something else—is as valuable in statistics as it is in other arts.

It would be wrong to leave the impression that people think of the four stages as they solve real-life problems, or that it would help them much if they did. But analyzing the process this way in retrospect is helpful in understanding how an inquiry progresses, and at what points statistics fits into it.

Applications of Statistics

So far we have discussed statistics at a general level. Now we pause to consider some of the kinds of practical and scientific problems to which statistics is applied.

Statistical methods have been increasingly used in business. One element common to all problems faced by business managers is the need to make decisions in the face of uncertainty; and, as we have seen, the essence of modern statistics lies in the development of general principles for dealing wisely with uncertainty. It is not surprising, then, that statistical methods are widely applicable in nearly all areas of managerial decisions. Applications are made in market and product research, investment policies, quality control of manufactured products, selection of personnel, the design of industrial experiments, economic forecasting, auditing, the selection of credit risks, and many others. The scientific management movement of this century has especially emphasized the need for collecting facts and interpreting them carefully, as has its currently popular offspring "operations research."

Governments have long collected and interpreted data concerning the State; for example, data about population, taxes, wealth, and foreign trade. In fact, the word *statistics* is derived from *state*. The first article of the Constitution of the United States provides that the government shall collect statistics—a decennial census to serve as a basis for representation of the states in Congress. Perusal of the *Statistical Abstract of the United States* will give an idea of the breadth and detail of statistical data currently compiled by the government: area and population; education; law enforcement; climate; labor force, employment, and earnings; elections; foreign commerce; transportation; and comparative international statistics are a few of its 34 major headings. The Department of Commerce has been a particularly important compiler of business statistics since World War I, and since World War II the President's Council of Economic Advisers has published data from a variety of sources on general economic conditions.[2]

Investigations in the social sciences have relied increasingly on statistical methods. The sample survey has supplied information at moderate cost on many topics, including incomes and savings; consumer anticipations about future expenditures; attitudes toward atomic energy, civil defense, public libraries, and international relations; voting, actual and intended; unemployment; and the effect of television on family life. Understanding of personality has been gained by statistical analysis of psychological tests and experiments. Attempts have been made to measure statistically the extent of monopoly in business at different times and thus the extent to which monopoly is increasing or decreasing. Archaeologists have used statistics in drawing inferences from excavated potsherds. The increasing use of mathematical "models" (that is, theories formulated in mathematical symbols) which attempt to explain social behavior has brought an increasing interest in statistical techniques by which the validity of these models can be tested.

The demands of research in certain biological sciences, notably anthropometry, agronomy, and genetics, brought forth a rebirth of statistics at the beginning of the twentieth century, and the use of statistical methods in this area continues to grow. The development of genetics, especially, has been intimately related to the development of statistics. Experiments about crop yields with different fertilizers and types of soil, or the growth of animals under different diets and environments, are frequently designed and analyzed according to statistical principles. Statistical methods also affect research in medicine and public health. The first large-scale, statistically well-designed medical experiment in the United States was done in 1952 to

[2] An excellent guide to government statistics is Philip M. Hauser and William R. Leonard (eds.), *Government Statistics for Business Use* (revised ed.; New York: John Wiley and Sons, Inc., 1956).

test the efficacy of gamma globulin as protection against polio-myelitis, though a statistically comparable experiment had been done in England in 1946 to test the efficacy of streptomycin in the treatment of tuberculosis.

The physical sciences, especially astronomy, geology, and physics, were among the fields in which statistical methods were first developed and applied (as early as the beginning of the nineteenth century), but until recently these sciences have not shared the twentieth century developments of statistics to the same extent as the biological and social sciences. Currently, however, the physical sciences seem to be making increasing use of statistics, especially in astronomy, chemistry, engineering, geology, meteorology, and certain branches of physics.

In the humanities—history, linguistics, literature, music, and philosophy, for example—the use of statistical tools is not common; but even in these fields statistics finds an increasing number of significant applications. A modern historian, for example, can use the evidence of attitude studies as well as more impressionistic data to characterize public opinion on, say, the question of isolationism in the United States just before World War II. An important historical question on which statistical evidence, even though fragmentary, has helped to give an answer, is whether the welfare of the working classes in England rose or fell during the industrial revolution of the late eighteenth and early nineteenth centuries. The power of statistics in resolving such an issue is illustrated by the fact that two authors who have done much to disseminate the view that the position of the working class greatly deteriorated during the early nineteenth century, admitted candidly toward the end of their lives that

> statisticians tell us that when they have put in order such data as they can find, they are satisfied that earnings increased and that men and women were less poor when this discontent was loud and active than they were when the eighteenth century was beginning to grow old in a silence like that of autumn. The evidence, of course, is scanty, and its interpretation not too simple, but this view is probably more or less correct.[3]

[3] J. L. and Barbara Hammond, *The Bleak Age* (revised ed.; London: Pelican Books, 1947), p. 15.

These allusions to statistical applications are not intended to be exhaustive, but simply suggestive of the diversity of applications of the underlying methods and ideas of statistics. Many more concrete illustrations will be given in later chapters. Statistics is a tool which can be used in attacking problems that arise in almost every field of empirical inquiry. While the details of the appropriate statistical techniques vary from one field to another and from one problem to another, it is important to recognize the basic similarity of approach. We hope to bring out this similarity and give insight into the scope and applicability of statistics by drawing illustrations from many fields.

But there is a deeper reason for stressing a broad range of applications. It is that the statistical approach, though universal in its underlying ideas, must be tailored to fit the peculiarities of each concrete problem to which it is applied. It is dangerous to apply statistics in cookbook style, using the same recipes over and over, without careful study of the ingredients of each new problem. A wide range of illustrations will, we hope, emphasize the need to begin from basic principles in attacking each new problem.

Our interest lies in statistical method. It is important to recognize, however, that statistics cannot be used to full advantage in the absence of good understanding of the subject to which it is applied. The statistician working in meteorology, for example, without a good understanding of meteorology is likely to produce technically competent trivia that contribute little to meteorology. Conversely, the meteorologist without a good understanding of statistics is likely to get bogged down in awkward, inefficient, and misdirected attempts to obtain evidence on important meteorological problems, and he is liable to fall into erroneous conclusions. The skill and knowledge of statistician and meteorologist must be blended. Sometimes the two abilities are combined in the same person, but more often the meteorologist consults with a statistician. Such collaboration relieves neither partner of the need to understand something of the other's field, but it does relieve each of the necessity of qualifying as an expert in two fields.

Factors Related to the Growth of Statistics

The great and continuing growth in the use of statistics can be explained by the economist's rubrics of demand and supply: The demand for statistics has increased, and so has the supply. Either increase, in the absence of a compensating decrease in the other, would bring about an increase in the use of statistics. The two increases together have magnified each other's effects.

Increased Demand for Statistics

The areas in which statistics is applied most are, as we have just seen, business, government, and science. The extraordinary growth of all three of these is one of the most distinctive features of the present century.

A striking, though indirect, reflection of the increasing importance of business is the fact that from 1910 to 1952, while the total civilian population of the continental United States was increasing by two-thirds (from 92 to 153 million), the farm population declined by more than a fifth (from 32 to 25 million); correspondingly, the percentage of the population classified as urban rose by nearly one-third (from 46 percent in 1910 to 59 percent in 1950)[4] and the percentage of workers engaged in nonfarm occupations rose by more than one-fourth (from 69 to 88).[5] Similarly, from 1929 to 1953, the number of business firms in operation increased half as much again as did the civilian population of the continental United States (by 39 and 26 percent, respectively).[6] The increased magnitude of business would alone account for a considerable increase in the need for statistics, but the need has been still further augmented by the increasing complexity of business, as firms have become

[4] *Statistical Abstract: 1955*, Tables 8, 9, and 16, pp. 13 and 24. The urban population from 1910 to 1950 was defined as "all persons living in incorporated places of 2,500 inhabitants or more and in other areas classified as urban under special rules relating to population size and density" (p. 2). Beginning with 1950, however, a new definition is used, differing chiefly by adding the residents of unincorporated places of 2,500 or more and of "the densely settled urban fringe . . . around cities of 50,000 or more." By this new definition, 64 percent of the population was urban in 1950.

[5] *Statistical Abstract: 1955*, Table 218, p. 185.

[6] *Statistical Abstract: 1955*, Table 577, p. 488, and Table 8, p. 13. The number of business firms in operation on June 30, 1953, was 4.2 million.

larger (in manufacturing the average number of employees per firm increased by more than a fourth between 1929 and 1953 [from 41 to 52]),[7] as government regulations and taxes have become more pervasive and complicated, as labor relations have become more involved, and as technology has advanced.

The increase in the magnitude of government is so often commented on that it will suffice here simply to cite two statistical facts: First, there were nearly two and one-half times as many government employees in 1953 as in 1919.[8] Second, the total expenditures of the federal government were about 150 times as great in 1953 as in 1900.[9] The increased complexity of government operations is illustrated by the facts that in 1910 there was no federal income tax and no social security program. Thus, government activities, even more than business activities, have increased in size and in complexity, thereby greatly increasing the demand for statistics.

The growth of scientific research has been equally dramatic. By 1954, funds used for research and development were three times as much as they had been only ten years earlier; universities were using more than five times as much, industry more than three times as much, and governments about twice as much.[10] Science too has become more complex, and this has resulted in a large increase in the demand for statistics in research.

Decreasing Costs of Statistics

The cost and the time required for summarizing and analyzing masses of data put a limit on the use of statistics. This limit has become progressively less restrictive because of technological improvements in processing numerical data. The de-

---

[7] *Statistical Abstract : 1954*, Tables 226 and 577, pp. 191 and 488.
[8] The numbers were 2.7 and 6.6 million, respectively. These include state and local, as well as federal employees, but not the armed forces, which numbered 3.6 million in 1953. *Statistical Abstract: 1955*, Tables 226 and 264, pp. 191 and 226.
[9] *Statistical Abstract: 1955*, Table 407, p. 349.
[10] Governments in 1954 were *providing*, in contrast with using, two and one-half times as much money for research as in 1944, industry was providing four times as much, and universities four times as much. *Statistical Abstract: 1955*, Table 593, p. 499. These figures somewhat exaggerate the growth of research activities, since they partly reflect price increases.

velopment of tabulating and computing machines has resulted in great savings of money and time, and, consequently, a marked impetus to the use of statistics. Recent developments, such as electronic calculators, have been spectacular. More conventional devices, such as desk calculators and card sorting and tabulating machines, have made it easy for scientists and administrators to complete statistical work that would have been too expensive and slow to undertake fifty years ago.

The development of statistical theory has also had the effect of reducing the costs of compilation of statistical data, especially by making it possible to base reliable conclusions on samples. At the beginning of the century, the idea of "taking a sample" had scant theoretical basis to serve as a guide to practice or to give confidence in the results. Great advances in the theory of sampling have occurred since that time; now procedures can be guided by these tried-and-proven theoretical developments. Two striking, though specialized, examples will illustrate the swiftness of these developments: (1) Sound techniques of *sampling human populations* when complete listings of individuals are not available have been developed almost entirely since 1935. As a result, estimates of unemployment are now prepared monthly with errors that almost surely are under 20 percent. Similarly, sampling methods make available important information from censuses long before the complete tabulations can be prepared and published. These are only two of the practical applications these sampling techniques have found; empirical research both in the social sciences and business is making increasing use of them in cases where, if it were necessary to have a complete listing of the individuals under study before selecting a sample, the cost and delay would be prohibitive. (2) Equally revolutionary developments have occurred, almost entirely since 1935, in the collection and analysis of many other kinds of data, especially through that branch of statistics known as the *design of experiments*.

Much of the basic progress in statistical theory of the past few decades can be attributed directly to a single individual, Sir Ronald Fisher (born 1890). As one writer puts it, "Fisher is the real giant in the development of the theory of statistics. His first paper was published in 1912, and his work continues unabated today. Although hundreds of scholars have contributed to the science of statistics, this one man must be credited with at least half of the essential and important developments as the theory now stands."[11] Fisher is not only the greatest figure in the history of statistics, but one of the greatest figures in the history of scientific method generally.

As rapid as the recent development of statistical theory has been, it would be wrong to give the impression that the current body of theory is complete or final. In spite of the rapid developments we have been outlining, the list of unsolved statistical problems is long, and statistical research today is more vigorous than ever before.

Conclusion

Statistics are numerical facts, but *statistics is* a body of methods for making decisions when there is uncertainty arising from the incompleteness or the instability of the information available. The decisions may be made either for the practical purpose of selecting a course of action or for the scientific purpose of gaining general knowledge.

Intelligent problem-solving, or scientific method, involves the observation of facts, the formulation of hypotheses describing the relations among the facts, the deduction from the hypotheses of things that must be true if the hypotheses are true, and the verification of these deductions by observing more facts. Statistics assists in planning the initial observations, in organizing them and formulating hypotheses from them, and in judging whether the new observations agree sufficiently well with the predictions from the hypotheses.

For the past two decades there has been a remarkable and sustained growth in the use of statistics. Partly, this is because business, government, and science, the three fields in which applications of statistics are most numerous and diverse, are

[11] Alexander McFarlane Mood, *Introduction to the Theory of Statistics* (New York: McGraw-Hill Book Company, Inc., 1950), p. 282.

growing in volume and complexity, both absolute and relative to other activities. Partly, too, it is because of a technological revolution in data handling, affecting especially computing and tabulating equipment, and a scientific revolution in statistical theories and techniques.

Chapter 2

Effective Uses of Statistics

Common Sense and Statistics

MOST OF us pass through two stages in our attitudes toward statistical conclusions. At first we tend to accept them, and the interpretations placed on them, uncritically. In discussion or argument, we wilt the first time somebody quotes statistics, or even asserts that he has seen some. But then we are misled so often by skillful talkers and writers who deceive us with correct facts that we come to distrust statistics entirely, and assert that "statistics can prove anything"—implying, of course, that statistics can prove nothing.

He who accepts statistics indiscriminately will often be duped unnecessarily. But he who distrusts statistics indiscriminately will often be ignorant unnecessarily. A main objective of this book is to show that there is an accessible alternative between blind gullibility and blind distrust. It is possible to interpret statistics skillfully. In fact, you can do it yourself. The art of interpretation need not be monopolized by statisticians, though, of course, technical statistical knowledge helps. This book represents an attempt to illustrate the fact that many important ideas of technical statistics can be conveyed to the nonstatistician without distortion or dilution.

Statistical interpretation depends not only on statistical ideas, but also on "ordinary" clear thinking. Clear thinking is not only indispensable in interpreting statistics, but is often sufficient even in the absence of specific statistical knowledge. Before we turn to the main stream of our exposition of statistical ideas, we shall devote this chapter and the next two to a series of statistical examples which can be interpreted reasonably well without any statistical background.

In the chapter after the next we will consider misuses of statistics, but in this chapter and the next one we will consider effective uses. In this chapter we will give quick sketches of successful applications of statistics in World War II, in business, in the social sciences, in the biological sciences, in the physical sciences, and in the humanities. Then in the next chapter we will take a closer, more detailed look at three examples, one each from the social, biological, and physical sciences.

One warning is needed before the examples are discussed. A receptive yet critical mind is essential. The rewards of open-minded skepticism are great, yet such skepticism is harder to apply than to advocate, especially when the problem in which statistical methods have been used is interesting. If one is interested in race relations in a community, in the effectiveness of an advertising campaign, or in the sexual habits of the population, it may seem tedious and pedantic to be critical about statistical methods. Statisticians are not much more immune to this attitude than anyone else, although they may be more consciously aware of it. One of the authors once recorded this reaction to an interesting book:

> When I first examined the volume, paying attention mostly to its fascinating substantive findings and scarcely at all to its methods, I was very favorably impressed indeed. When I diverted my attention to the general methods I began to note shortcomings; but I felt that these were technicalities—mere blemishes on the surface of the monument, which might modify some of the findings in detail but surely would not affect the broad conclusions. After all, many of [the] figures would still be important and interesting even if we had to allow for an error factor as large as two or even three. But when I spent some time studying the statistical methods in detail, I realized that my confidence in the basic significance of the findings cannot be securely buttressed by factual material included in the voluem. In fact, it now seems to me that the inadequacies in the statistics are such that it is impossible to say that the book has much value beyond its role in opening a broad and important field.[1]

Even in the successful examples that follow, one should note potential flaws, and consider what effect they might have on special applications of the findings.

[1] W. Allen Wallis, "The Statistics of the Kinsey Report," *Journal of the American Statistical Association*, Vol. 44 (1949), p. 466.

Some Uses of Statistics in World War II

EXAMPLE 31A AIRCRAFT LOSSES IN RELATION TO TIME SINCE OVERHAUL

In order to minimize flying time lost for overhauling engines, and at the same time avoid plane losses that overhauling could have prevented, a study was made of the relation between aircraft losses and flying time since the last overhaul. Contrary to expectation, it was found that the number of planes lost decreased as the time since overhaul increased; that is, the risk of failure was greatest right after overhaul, and declined steadily until the next overhaul, when it increased again. This result led to a great extension of the amount of flying time between overhauls. It also led to an investigation and reorganization of the overhauling system, so that overhauling made the planes less rather than more likely to fail. This improvement in the overhauling system illustrates again the point about serendipity that we made in Example 18 (Theft of Finished Product), that unanticipated by-products of a systematic statistical study may be at least as useful as the specific objectives. This important study required little more than intelligent and careful collection of data, and their proper organization and interpretation.

EXAMPLE 31B MERCHANT SHIP LOSSES IN RELATION TO CONVOY SIZES

In 1942–1943, large merchant ship losses by submarine attack led to a study of the relation between the number of ships lost and the size of the convoy. Data on losses for various sizes of convoy revealed that there was no tendency for the number of ships lost to vary with the size of the convoy, though, of course, there was considerable variation in losses even for convoys of a given size. Since the *number* of ships lost did not increase with the size of the convoy, the size of the convoys was increased to reduce the *percentage* loss. One possible explanation of this independence of loss and convoy size is a constant attack potential of a submarine group.

EXAMPLE 32A   ARMY USE OF SAMPLING INSPECTION OF MASS-PRODUCED ITEMS

Because of the tremendous amount of work required for complete inspection of mass-produced items, the Army, with the guidance of Bell Telephone System statisticians, introduced sampling inspection plans. Under such plans, only a small part of an entire lot (perhaps only 100 out of 5,000), is inspected in order to determine whether the entire lot should be accepted or rejected. Unless such sampling is statistically sound it may be worse than useless, but if properly done it is usually superior to inspecting each item, because the few items can be inspected more accurately; and, of course, it is far cheaper.

EXAMPLE 32B   OPA SAMPLE STUDIES OF THE INVENTORIES

During World War II, the Office of Price Administration attempted to take complete inventories of tires in the hands of dealers. Later, a number of dealers were selected on a statistical basis and the complete inventory was estimated on the basis of this relatively small group of dealers. Not only was a nuisance eliminated for many dealers, but the figures proved more accurate than the complete inventory previously attempted. The increase in accuracy was due to the fact that there was a huge number of nonresponses in the "complete" counts, which were made by mailed questionnaires, but in the sample it was possible by energetically following up to keep nonresponses at a low level. Those who failed to respond without follow-up proved to be quite different from those who responded readily.

EXAMPLE 32C   ESTIMATES OF ENEMY OUTPUT

During the war, German industrial output and capacity were estimated by British and American statisticians from the manufacturing serial numbers and other marks on captured equipment. According to checks after the war, many of these estimates were quite as good as the estimates made by the Germans themselves. They were, moreover, available substantially sooner than the estimates of the Germans, since the Germans waited for complete coverage, whereas the British and Americans were forced to rely on sampling methods. The Germans never did know their total production figures for V-2 missiles, most of which were produced towards the end of the war, while the British and American estimates subsequent to the firing of the first missile, were found by special studies after the war to have been quite accurate.

EXAMPLE 33A   RELATION BETWEEN TRAINING
AND BOMBING ACCURACY

The question of the most fruitful division of flying time between training and bombing missions for B-29 airplanes operating from the Marianas was resolved by a statistical study of the relation between training time and accuracy of bombing. It was found that with an increase of training time from four percent to 10 percent of the available flying time, the number of bombs on the target doubled.

Some Uses of Statistics in Business

EXAMPLE 33B   FITTING A NEW PRODUCT
TO CONSUMER TASTES

A flour manufacturer wanted to bring out a new pie crust mix. The proposed mix was put in plain packages with only an identifying letter, and the same was done with the pie crust mix of the leading competitor. Each of 75 families was given a package of each of the two mixes and asked to fill out a questionnaire comparing the two brands for taste, texture, and other qualities. The competitor's product was preferred by most of the 75 families. The flour company therefore revised the formula of its mix. The revised formula was tested on another group of 75 families and found still to be less preferred than the competitor's mix. The same thing was repeated several more times. Meanwhile, another competitor introduced a new mix which gained preference over all others. The company made similar tests of its proposed formulas against this new competitor. Finally, a mix was developed that seemed to be preferred to all competitive products. When it was marketed it proved highly successful.

EXAMPLE 34A   ESTIMATING SALES BY DEALERS TO CONSUMERS

A manufacturer of household appliances wanted to know the current rate of sales of his appliances to consumers. The only information readily available was the rate of factory shipments to distributors and dealers. There were more than ten thousand distributors and dealers, however, and only vague, qualitative reports were available on the rate at which dealers sold appliances to the public. It may take as long as two or three months before changes in dealers' sales are reflected in changes in orders to the factory. A group of about 100 dealers was chosen, and each was asked to fill in a monthly report of his sales. Changes in the rate of sales to consumers were estimated from these reports. These data proved extremely valuable later when orders to the factory declined sharply. It was found from the reports of the sample dealers that the decline was due to reductions of inventories by dealers, rather than to any marked drop in retail sales.

EXAMPLE 34B   VALUATION OF PLANT AND EQUIPMENT

A telephone company wished to determine the value of its capital equipment, such as poles, cables, batteries, tools, buildings, and so on. It had lists of these things, and knew the original and the replacement costs, but the question was what percent of their useful life still remained. For many items this is costly to estimate, perhaps because the item is in a remote area or an inaccessible position, or because the evaluation requires highly paid experts or requires dismantling equipment and disrupting service. Furthermore, the number of items was large. Instead of examining every piece of equipment, which would have been prohibitively slow and costly, the company chose a relatively small sample of each kind of equipment according to statistical principles. The average condition of each kind of item was then estimated from the samples, and an allowance was made for possible error due to sampling. The results were as useful as if a full examination had been made.

EXAMPLE 35A QUALITY ASSURANCE

A manufacturer of electric fuzes wished to guarantee that his product met specified standards of quality. Quality was defined in terms of the time required for the fuze to break the circuit, when a certain current was applied. To find out whether a fuze met the standard of quality, the item had to be destroyed. A statistically designed sampling plan, similar to those mentioned in Example 32A (Army Use of Sampling Inspection), enabled the manufacturer to draw reliable inferences about the quality of an entire lot from the information obtained by testing relatively few items.

EXAMPLE 35B EXPERIMENTING ON A MANUFACTURING PROCESS

A manufacturer made a very expensive product which, like the one in the preceding example, could be tested only by destroying the product. His engineers wanted to know the effect on the final product of changes in six variable components of the manufacturing process, such as amounts and kinds of various ingredients, and the length of time certain operations were continued. The traditional approach would have been to vary one component at a time and observe the effect. In this case, however, that would have been too costly, for it was necessary that each component be tried at five different "levels"; for example, each of five different concentrations of one of the ingredients had to be tried. If, say, five units of the product were to be made and tested at each of five levels of one of the components, 25 units of the final product would have had to be specially made and tested, and for six components a total of 150 units would have been required. A statistician, however, devised a plan for the experiment which permitted the simultaneous evaluation of the effects of all six components on the basis of testing only 25 units of the final product. It was possible to determine whether each component had an effect on quality beyond that which might be reasonably attributed to chance or random variations in the product. For those components which had a significant effect, it was possible to estimate what the effect was, so that an approximation to the "optimum" level of the component could be used in the manufacturing process. This is not to say that everything that could have been learned from a full 150 tests was learned from 25; but the specific questions of most importance were answered reliably enough from 25 tests.

EXAMPLE 36A ESTIMATING SALES OF DIFFERENT STYLES

In order to place orders with manufacturers, a mail order company wanted to predict how total sales of a certain garment would be divided among individual styles. To a randomly selected sample of catalog receivers, the company mailed, in advance, a special booklet made up of the catalog pages describing the various styles of the garment. Actual orders from recipients of the special booklet were then tabulated, and predictions were made on the basis of these orders. These predictions were found, at the end of the sales season, to have been substantially more accurate than the predictions of experienced buyers.

EXAMPLE 36B SEASONAL PATTERNS OF ACCIDENT RISKS

A personnel manager wanted to find out the times during the year when the largest number of accidents occurred in his plant. With this information he hoped to be able to give safety instruction when the need for it was greatest. A statistical study of the accident records for this plant showed that, while there were variations among the months in the number of accidents, these variations were no greater than might reasonably be expected by chance alone. Thus, there was no best season for safety indoctrination, and the decision could be based on other grounds.

EXAMPLE 36C USE OF RESERVED FACILITIES

A large firm arranges frequent educational programs for its 600 supervisory employees from foremen up. Each program includes discussion, and 30 is considered the best number of participants. Each program is therefore given 20 times, twice each morning and afternoon, Monday through Friday. Each employee is assigned to a session, but is free to come to any other session instead of the assigned one, if he feels that he ought not to leave his work at the time assigned. Originally, 30 employees were assigned to each session. Records of actual attendance showed fewer than 10 at some sessions and more than 60 at others. They also showed considerable uniformity for corresponding sessions of different weeks. The number of assignments was therefore varied, 90 being assigned to a session that had averaged 10 in attendance on the assumption that one-third of those assigned to that session would attend it, 15 to a session that had averaged 60, and so forth. Thereafter, actual attendance was seldom less than 25 or more than 33.

Hotels, airlines, physicians, restaurants, and others who make reservations that are subject to cancellation by clients sometimes use a variant of this system, but the problem is more complicated when there is an inflexible upper limit on the number who can be accommodated. Another variant was used with some success during World War II by the cafeterias in the Pentagon. They posted charts showing the average length of the line at various times. The charts revealed considerable variation in the wait encountered at times fairly close together. After the charts were posted these variations were appreciably reduced, as some people with control of their own lunch periods avoided the times of longest delay.

Recently, problems of this general kind have been considered extensively under the name "queuing theory." (In England, a waiting line is called a queue.)

Some Uses of Statistics in the Social Sciences

EXAMPLE 37A   CONTENT ANALYSIS

A political scientist studied British attitudes toward the United States during the period 1946–1950, insofar as these attitudes were expressed in newspapers and records of parliamentary debates. By a technique known as "content analysis," which consists of finding the relative frequency of appearance of different "themes," he hoped to describe British attitudes and to detect changes through time. A sampling scheme was devised whereby instead of reading all of the issues of the leading

British papers during the period of interest, he read only a selected sample.

EXAMPLE 38A CONSUMER FINANCES

For several years the Federal Reserve Board has sponsored at least one survey a year in order to determine, among other things, basic facts about income, savings, and holdings of liquid assets by individuals. A sample consisting of about 3,000 families has been used in this study. One interesting by-product of this work has been some indication that consumer plans for purchasing durable goods may be helpful in forecasting general business conditions.

EXAMPLE 38B SUCCESS IN COLLEGE

Numerous studies have demonstrated considerable correlation between scholastic performance in high school and performance in college. There is also some, but less, correlation with performance on entrance (or "aptitude") examinations. Findings from such studies have been used to predict "success" of students applying for entrance to college, on the basis of their high school performance and entrance examinations.

EXAMPLE 38C PUBLIC OPINION

Impressions about public attitudes on important issues are often ambiguous and unreliable. Carefully designed statistical studies can usually give more accurate pictures. An interesting example is the results of the following two questions asked in Belgium in 1948.[2]

	Yes	No	No Opinion
		(in percent)	
Do you believe that the American government sincerely wants peace?	68.5	13.9	17.6
Do you believe that the government of the U.S.S.R. sincerely wants peace?	18.0	60.2	21.8

[2] Institut Universitaire d'Information sociale et économique (Centre belge pour l'étude de l'opinion et des marchés), *Cinq Années de Sondages* (Brussels: 1950), p. 62. The wording of the question is here translated from the French.

EXAMPLE 39   HOUSING SUPPLY

The following use of statistical reasoning appeared as part of a discussion of rent controls and the postwar housing shortage:

> The present housing shortage appears so acute, in the light of the moderate increase in population and the actual increase in housing facilities since 1940, that most people are at a loss for a general explanation. Rather they refer to the rapid growth of some cities—but all cities have serious shortages. Or they refer to many marriages and the rise of birth rates—but these numbers are rarely measured, or compared with housing facilities.
>
> Actually the supply of housing has about kept pace with the growth of the civilian nonfarm population, as the following estimates based on government data show:

| Date | NONFARM | | |
	Occupied Dwelling Units	Civilian Population	Persons per Occupied Dwelling Unit
June 30, 1940	27.9 million	101 million	3.6
June 30, 1944	30.6 million	101 million	3.3
End of Demobilization (Spring 1946)	More than 31.3 million	About 111 million	Less than 3.6

> Certain areas will be more crowded in a physical sense than in 1940, and others less crowded, but the broad fact stands out that the number of people to be housed and the number of families have increased by about 10 percent, and the number of dwelling units has also increased about 10 percent.[3]

Thus, the authors found that an explanation of the unavailability of housing had to be sought elsewhere than in a physical shortage.

[3] Milton Friedman and George J. Stigler, *Roofs or Ceilings? The Current Housing Problem* (Irvington-on-Hudson, New York: Foundation for Economic Education, Inc., 1946), pp. 17–18.

Some Uses of Statistics in the Biological Sciences

EXAMPLE 40A   HEIGHTS OF PARENTS AND CHILDREN

By recording the heights of parents and children and group-these data by the height of one parent, it has been found that for every inch by which the parent's height exceeds (or falls below) the average for adults of the same sex and generation, the average of the children's heights, when grown, exceeds (or falls short of) the average for their sex and generation by about half as many inches. If data are grouped by the heights of both parents, the children's average is about four-fifths as far from the general mean, and in the same direction, as the parents'.

EXAMPLE 40B   MENDELIAN HEREDITY

Gregor Mendel discovered the foundations of the modern science of genetics about a century ago, by methods that were essentially statistical. Mendel, working with garden peas, noted the characteristics of the parents and counted the number of offspring having various characteristics. The regularities he observed led to the formulation of his theories.

EXAMPLE 40C   ANIMAL POPULATIONS

To determine the number of mice in a field or the number of fish in a lake, biologists catch a sample, count them, mark them (often with metal tags), and release them. Later they catch another sample. If, say, ten percent of those in the second sample are marked, they can infer that the total population is about ten times as large as the first (tagged) sample. Various elaborations are necessary to allow for special circumstances (such as that in a large woods or lake, a mouse or fish is likely to remain in a certain general area), to improve the estimates, and to calculate an allowance for error in the estimate.

Some Uses of Statistics in the Physical Sciences

EXAMPLE 40D   DIVISION OF THE TERTIARY ROCKS

Charles Lyell, the geologist, published the three volumes of his celebrated *Principles of Geology* in 1830, 1832, and 1833.

Geologists prior to Lyell had recognized the sequences of strata which we know as Primary and Secondary, using in the first place the regularity of order of superposition in the same locality. They observed, too, that particular components of these formations could be recognized, though far apart, by their characteristic fossils. They could not by these means recognize or establish the order among Tertiary rocks, for, in the part of the world then accessible, these occur in patches, and not over wide areas overlying one another. Lyell determined the order and assigned to the successive rock masses the names they now bear by a purely statistical argument. A rich group of strata might yield so many as 1,000 recognizable fossil species, mostly marine molluscs. A certain number of these might be still living in the seas of some part of the world, or at least be morphologically indistinguishable from such a living species. . . .

With the aid of the eminent French conchologist M. Deshayes, Lyell proceeded to list the identified fossils occurring in one or more strata, and to ascertain the proportions now living. To a Sicilian group with 96 percent surviving he gave, later, the name of Pleistocene (mostly recent). Some sub-appenine Italian rocks, and the English Crag with about 40 percent of survivors, were called Pliocene (majority recent). Forty percent may seem to be a poor sort of majority, but no doubt scrutiny of the identifications continued after the name was first bestowed, and the separation of the Pleistocene must have further lowered the proportion of the remainder. The Miocene, meaning "minority recent," had 18 percent, and the Eocene, "the dawn of the recent," only 3 or 4 percent of living species. Not only did Lyell immortalize these statistical estimates in the names still used for the great divisions of the Tertiary Series, but in an Appendix in his third volume he occupies no less than 56 pages with details of the classification of each particular form, and of the calculations based on the numbers counted. There can be no doubt that, at the time, the whole process, and its results, gave to Lyell the keenest intellectual satisfaction.[4]

EXAMPLE 41   RADIOCARBON DATING

Radiocarbon, or Carbon 14, is present in all living things. While things are living, the quantity of radiocarbon is proportional to the quantity of nonradiocarbon, or Carbon 12. After

[4] Sir Ronald Fisher, "The Expansion of Statistics" (Inaugural Address as President of the Royal Statistical Society), *Journal of the Royal Statistical Society, Series A (General)*, Vol. 96 (1953), pp. 1–6.

death, the quantity of nonradiocarbon remains stable, but the radiocarbon disintegrates. The ratio of radio- to nonradiocarbon indicates, therefore, how long a specimen has been dead.

There are standard chemical methods of determining the amount of nonradiocarbon in a specimen. The radiocarbon emits small particles that can be detected by special counting devices, such as the Geiger counters and scintillometers used in uranium prospecting or in measuring the radioactive fallout after atomic explosions. The average rate at which these particles are emitted is proportional to the amount of radiocarbon present, and hence provides a means of measuring the amount of radiocarbon present, but the actual emission at any given time is a matter of chance. From counts of the number of particles emitted in a given period it is possible, by statistical methods, to determine the average rate, and to determine the necessary allowance for uncertainty in this average due to the chance character of the emissions. The average rate of emission indicates the amount of radiocarbon present, and the ratio of this to the amount of nonradiocarbon indicates the age of the specimen. Allowances for uncertainty in the age are calculated from the corresponding allowances for uncertainty about the average rate of emission of particles from the radiocarbon.

Radiocarbon dating has become a standard means of dating ancient materials such as textiles, leather, and wood charcoal from campfires; it has revolutionized the dating of archaeological objects.

Some Uses of Statistics in the Humanities

EXAMPLE 42 LINGUISTIC DATING

A method statistically similar to Carbon 14 dating, and in fact suggested by it, has been used in linguistics. In place of Carbon 14, it uses a list of two to three hundred concepts for which there are words in virtually all languages. By studying languages which are known to be descendants of a common language, for which the date of separation is known, and for which there are writings at various known dates following the separation, it has been found that after separation the number of common words tends to diminish by about 20 percent per 500 years. Thus, after 500 years, about 80 percent of the words are still the same, after 1,000 years, about 64 percent, and so forth. Knowing this, it is possible to calculate, from the number of words they have in common, when two related languages separated. This method, however, is not so firmly established and widely used as radiocarbon dating.

EXAMPLE 43 LITERARY STYLE

Statistical studies of the length of sentences, the relative frequency of various parts of speech, the frequency of use of individual words, and the frequency of various word sequences have been used to help answer such disputed questions as whether a given author wrote a certain work, whether a work came early or late in an author's career, and what portions of joint works were written by the respective authors.

Conclusion

Statistical methods are used effectively in the most diverse subjects, ranging from minor business and personal decisions to abstruse questions of pure science and scholarship.

Brief illustrations serve to indicate the range of applicability of statistics, but they can give only the barest hints about the way statistics enters into these applications. Statistics, when used effectively, becomes so intertwined in the whole fabric of the subject to which it is applied as to be an integral part of it. Full appreciation of the ways in which statistics enters into an investigation requires, therefore, a detailed analysis of the subject matter and of all the methods brought to bear on it.

Chapter 3

Psychoses, Vitamins, and Rain-Making: Three Extended Examples

THIS CHAPTER is devoted to rather detailed examinations of three successful statistical studies, one each in the social, biological, and physical sciences. The first example, on long-term trends in the frequency of mental disease, involves a historical study in which the investigators had to rely on existing data and records, whereas the second example, on the effect of vitamins B and C on human endurance under severe physical stress in extreme cold, and the third, on making rain by "seeding" clouds, involve experiments arranged by the investigators for their specific purposes. The second and third examples, though completely different in subject matter, are in many respects similar statistically.

The purposes of presenting these three examples are to dispel any aura of magic that may have resulted from the brief summaries in the preceding chapter, to present a glimpse of the inner "works" of a statistical investigation, to impart a feel for the necessity of caution, judgment, and detailed information in drawing conclusions from even the best research, and to indicate the extent to which the over-all soundness of an investigation depends on care and skill with a large number of details. In these examples, therefore, instead of omitting details and focusing on the major methods and findings, we shall give particular attention to details, though it will be impractical to recapitulate the original studies in full detail.

Long-Term Trends in the Frequency of Psychoses

Purpose of Study. Mental health is a matter of growing concern. War, depression, urbanization, industrialization, competition, and the breakdown of family and community ties: these and many other aspects of the increased complexity and insecurity of modern civilization are said to have aggravated the problem. To determine how much increase there has been in the frequency of mental disease over the past century was the purpose of a careful investigation by Herbert Goldhamer, a sociologist, and Andrew W. Marshall, a statistician.[1]

Such a purpose is too broad for a single inquiry and too vague for a systematic one. Goldhamer and Marshall, therefore, proceeded to narrow and to define their objectives. The result is a study which leads to reliable and specific conclusions about a single facet of the broader problem. Studies of other facets, made equally carefully but perhaps by other investigators in other times or countries, each building on its predecessors, will eventually cumulate into an understanding of the broader problem. Even this broader problem is, as Goldhamer and Marshall emphasize by their title, only a facet of the still more fundamental question of the relation between the psychogenic psychoses (serious mental disorders which apparently result from mental influences, rather than from physiological influences, injuries, or other causes) and the characteristics of contemporary social existence, particularly the characteristics associated with the growth of "civilization," especially its increased personal responsibility and freedom.

First, Goldhamer and Marshall specified more explicitly what they mean by *frequency*. Obviously they are not interested in the total number of cases, since this would mainly reflect the fact that the population of the United States is about six times as large as it was a century ago. Nor are they interested in the number of cases of mental disease per capita of population. True, this would allow for changes in the total population, but it would not allow for the fact that the frequency of mental disease varies with age, and now a larger proportion of the population is at the ages most susceptible to mental disease. This change in the age distribution of the population would cause an increase in the per capita rate of mental disease even if the rate of onset for each specific age group were unchanged.

[1] Herbert Goldhamer and Andrew W. Marshall, *Psychosis and Civilization: Two Studies in the Frequency of Mental Disease* (Glencoe, Illinois: Free Press, 1953).

What they decided to use was not a single frequency figure but a set of *age-specific rates*. These show, separately for each age group, the number of cases of mental disease per capita of population of that age group.

Furthermore, mental disease is too broad and vague a concept, so the authors narrowed it to include only the *major psychoses*. These are characterized by "behavior, such as extreme agitation, excitement, deep depression, delusions, hallucinations, suicidal and homicidal acts" that is "sufficiently recognizable as insanity, irrespective of the classificatory terminology used" so that we may be confident that such cases would be diagnosed as mentally diseased either a century ago or today. Some of the lesser forms of disturbance, such as neuroses, "nervous breakdowns," and "maladjusted personalities," are not clearly classifiable as mental disease, have been considered such only comparatively recently, and may be classified differently by different diagnosticians even contemporaneously. Thus, data that included them probably could not be found at all, and if found would be virtually worthless for comparisons among widely separated times.

In order to obtain data on the number of cases of psychosis, it was necessary to narrow the study still further, covering *hospital admissions*. Not all admissions to mental hospitals, but only *first admissions*, are relevant, for the authors wish their figures to show the rate of onset of mental disease.

Whether hospital first admissions for psychosis are an adequate index of the number of cases of psychosis in the population requires consideration. It may be that the relation between hospital first admissions and the actual frequency of mental disease has changed in the course of time. It seems plausible, for example, that the proportion of the afflicted population hospitalized is larger now than a century ago and that hospitalization follows sooner after the onset of the disease. These differences seem plausible because there has been an increase in the number of beds available, relative to the need for them, and also because the proportion of cases hospitalized is greater for cases near hospitals and, now that the population is more urban (the proportion urban rose from 15 to 60 percent between 1850

and 1950), more people are close to hospitals. This matter of
the relation between hospital first admissions and the total incidence of mental disease is one that, as we shall see, is kept
constantly in mind by the authors in their analysis and interpretation.

Narrowing the study to hospital first admissions is dictated
not by the desire for definiteness and precision, but by the avail-
ability of data. A shift has been made from what would be
studied ideally, namely, rates of first onset for the whole population, to what can be studied practically, namely, rates of first
admission to hospitals. Such shifts are commonly necessary in
research. They require especially good judgment, for neither a
precise study of irrelevant trivialities nor a meaningless study
of the central issue is of any value. Good researchers must
balance tenacious adherence to strategic objectives against at-
tacks on targets of opportunity. As a matter of fact, this whole
investigation is an excellent example of the role of serendipity,
the art of successfully exploiting good things encountered accidentally while searching for something else, in scientific progress. Goldhamer and Marshall intended to study variations in
the frequency of mental disease among different groups of the
contemporary population. While searching in the Library of
Congress for pertinent data, they came across data which, they
recognized, made possible a study of long-term trends, something they had presumed would be impossible.

A final step in defining the specific objectives, also dictated
by expediency, was to confine the study to the state of Massachusetts:

> Massachusetts was chosen as the state of inquiry because its
> facilities for the care of the mentally ill during the last half of
> the 19th century were, despite their obvious limitations, more
> advanced than those of most other states. The relatively small
> size of Massachusetts is also favorable to our inquiry, since it
> diminishes the mean distance of the population from a mental
> hospital. It has been well known for some time now that the
> tendency, especially in the past, to hospitalize the mentally ill
> is inverse to their distance from a mental hospital. Massachu-
> setts further recommends itself for study during the latter half
> of the 19th century because of the work of such leaders in

institutional psychiatry as Dr. Edward Jarvis and Dr. Pliny Earle. Their studies and reports, together with the documents and reports of official state agencies and hospitals, made feasible an investigation which one might have supposed quite impossible at this late date.[2]

Clearly the authors would not be greatly interested in a single state unless they felt that conclusions might be drawn which would apply reasonably well to some larger area, such as the entire United States. In short, they were faced with the problem of drawing conclusions about a *population* or *universe* (here the entire United States) on the basis of a *sample* from that universe (here Massachusetts). Sampling concepts pervade modern statistics and will be discussed at length in this book. We will see that when the element of *randomness* is introduced in the selection of the sample, effective techniques are available for drawing inferences about the population from which the sample was selected. In the present example, however, randomization was not possible, since Massachusetts was the only state for which accurate and meaningful information was available. It was necessary, therefore, to rely on expert judgment to decide how far the results of the sample might be generalized—to decide, that is, how far what is true of Massachusetts is approximately true of the entire United States or some other large region of importance. Goldhamer and Marshall feel that their findings do apply more widely, and they present some supporting evidence to be described later in this chapter. In so deciding, they were acting very much like a surgeon at a private clinic who decides that a new operative technique is successful even though he knows that the patients on whom he has tried it are not randomly selected but are definitely atypical in many respects, for example, income. He may judge, however, that in respect to the operation, his patients are similar to patients of the same age and sex of other income groups. He would want, of course, to examine every bit of evidence that bears on his judgment. While Goldhamer and Marshall have reason to believe that Massachusetts fairly reflects major trends in the incidence of the

[2] Goldhamer and Marshall, *op. cit.*, pp. 25–26. Supporting footnote omitted.

psychoses, they undoubtedly would have preferred more evidence on this point than they had.

The authors summarize their purpose, then, as follows:

> The immediate aim of this report is to establish acceptable estimates of age-specific first admission rates to institutions caring for the mentally ill in Massachusetts for the years 1840 to 1885 in order to compare these rates with those of the contemporary period. No antiquarian zeal or historical interest has moved us to engage in this laborious task.* Our interest is in providing a more adequate test than is now available of contending views concerning the incidence of the major mental disorders in our own day and in an earlier period. We assume that a more adequate test of these beliefs will throw light on the validity of contentions concerning the psychologically pathic effects of contemporary social existence. . . .

What Was Known Already? Goldhamer and Marshall mention and discuss briefly eight previous studies related to their problem. Most of these suggest that the incidence of mental illness has *not* increased; only one seems to conclude that an increase has taken place. Goldhamer and Marshall did not consider these previous studies to be decisive, chiefly because the time periods investigated did not extend very far into the past, some covering periods as short as a decade and only one extending back into the 19th century, and most of the studies (including the one covering the longest period of time) did not make adequate adjustments for the changing age-composition of the population.

In addition, the Goldhamer-Marshall volume is influenced by their preliminary study of their own data. This suggested strongly that there has in fact been no increase—that the 19th century rates of mental disease were fully as high as those of today, quite contrary to what we suggested in introducing this example. This tentative finding influenced their analysis, in that wherever questions of judgment, of estimating, or of selection

---

* Nor were we concerned to show that a judicious use of documents usually reserved for historical study can extend the horizons of comparative statistical social analysis to phenomena other than those of a demographic and economic character. Yet the present study does show that the past is not always as irrecoverable, statistically speaking, as is sometimes assumed. [Footnote in original. Goldhamer and Marshall, *op. cit.*, p. 21.]

among different figures arose they were careful that if they erred it would be in the direction of overstating the amount of increase. If, despite such bending over backward, it appeared that there had been no increase, they could be fairly confident that this conclusion was justified. Of course, if there had actually been a decrease they might fail to establish it this way; but for their purpose of determining whether contemporary social existence has caused an increase in the psychoses it is sufficient if they show either that there has been an increase or that there has not. Put another way, they were trying to decide between two actions, to investigate further the notion that the pace of modern life is a cause of psychosis, or to drop this notion and seek other causes. The first decision would be right if there has been an increase in the psychoses, the second if there has been no change or if there has been a decrease. In statistical terms that are self-explanatory in this context, the null hypothesis was that there has been no change, and it was tested against the one-sided alternative hypothesis that there has been an increase.

How the Data Were Obtained. Though the basic records were surprisingly good for so early a period, it was nevertheless a lengthy and demanding task to derive from them the age-specific rates of first admission. In the authors' own words:

> It is not to be denied that even the presentation of *first* admission rates for so early a period is in itself a novelty and that the attempt to carry this to a further stage of refinement by calculating *age-specific* first admission rates may seem extremely hazardous. However, we shall present quite fully the sources of our data and all assumptions involved in their utilization. The reader will be in a position to judge for himself whether any of our assumptions were or were not warranted. . . .[3]

While it is desirable to limit statistical investigations to things that can actually be observed, it is sometimes necessary to introduce assumptions which cannot themselves be tested completely. An author's willingness to tell exactly what he did, without vague general assertions as to the high quality of the work, is

[3] Goldhamer and Marshall, *op. cit.*, p. 26.

presumptive evidence of the soundness of the work and of the integrity of the author.[4]

Goldhamer and Marshall present eleven pages in which the data, method of collection, and assumptions are discussed in detail. A few highpoints of this discussion will be summarized. Before you read this summary, you might well pause to think of possible questions and objections. Then, as you read, notice how many of your questions are answered by the authors and how many unanswered, and how many questions the authors thought of that did not occur to you.

(a) Almost complete records had survived for all the mental hospitals and town almshouses in Massachusetts for the period 1840 to 1885.

As for the general accuracy of the admission bookkeeping and reporting of the hospital and state reports of this period, we can only say that intensive reading, analysis and cross-checking of them have given us the highest esteem for the thoroughness and integrity of the hospital admission data they present. The system of financial accounting for the support of state and town paupers and for receipts from privately paying patients made accurate records imperative. In the earlier years especially, the hospital "paper work" was looked after by the medical superintendent himself. One is impressed by the almost obsessive detail and thoroughness of the statistical summaries of some of the reporters. In a number of respects the early Massachusetts hospital reports are more illuminating than those produced today.[5]

[4] One recent investigator asserts boldly in the introductory chapter to his book that "It is a fact that no number reported in this study is exact." After a discussion of the sources of error and the measures taken to cope with them, he goes so far as to extend an invitation to bona fide scholars or journalists to come to the office where the documents are filed and to read some or all of them at their leisure. Samuel A. Stouffer, *Communism, Conformity, and Civil Liberties: A Cross-section of the Nation Speaks Its Mind* (Garden City, N. Y.: Doubleday and Company, Inc., 1955), pp. 23–25.

This attitude is in sharp contrast to that of a research organization which refuses to release its original data on the ground that they might be misinterpreted. They might. There will be somebody to misuse virtually any set of data: in fact, wherever there is freedom of speech there will be (as advocates of freedom recognize) erroneous and mendacious statements. There will be much more misinterpretation, error, and prevarication, however, and it will gain wider currency and survive longer, when there is protection from independent inquiry and analysis.

[5] Goldhamer and Marshall, *op. cit.*, pp. 31–32.

(b) There were some gaps in the hospital records. The authors present a detailed table which shows the "Number of first admissions to institutions caring for the insane, Massachusetts, by 5-year periods, 1840 to 1884, and 1885."[6] (The quotation marks enclose the title of the table itself; the clarity, completeness, and brevity of this title provide a good model for statistical practice.) The table is accompanied by a page of footnotes which, together with textual discussion, describes exactly how the data were obtained for each hospital and how allowances were made for incomplete information. For example, for McLean Hospital, a private mental hospital in Boston, the exact number of first admissions was available from reports for the years 1868 to 1885, but for the years 1840 to 1867 only total admissions were available. The ratio of first to total admissions during 1868–85 was 0.70. It was assumed that the same ratio prevailed in the earlier period, and an estimate was made on the basis of this assumption. For some of the other hospitals, data on first admissions were available for the entire period 1840–85.

Another example (the comment in brackets is ours):

> Beginning with 1870 our table shows a small number of admissions from other private hospitals. [The number of patients admitted to these hospitals in 1870–74 was only 114, about three-tenths of one percent of the total.] We know that there were two or three private mental "hospitals" in earlier years, but the number of patients (about 25) that they housed at that time is negligible. Since we were unable to secure further data, we have no entries prior to 1870 for this class of small private hospitals. . . .[7]

This is one of the points where the authors used judgment in such a way that the error, if any, would tend to favor the hypothesis that mental illness had really increased.

(c) The care needed to obtain meaningful numbers is illustrated in the derivation of the age distribution of first admissions (comments in brackets are ours):

[6] *Ibid.*, pp. 28–29.
[7] *Ibid.*, p. 27.

In order to arrive at age-specific first admission rates we required the age distribution of patients admitted for the first time to an institution. For the entire period, 1840 to 1885, wherever first admission data were not available it was possible to secure from the official reports of the state mental hospitals the age distribution of their *total* admissions. For the South Boston Hospital [a private hospital] we were able to secure the age distribution of total admissions only for the years 1850–54 and 1860–64. The age distribution for these two periods was very close to that provided by the state hospital reports for the corresponding years; we have assumed that the age distribution of the South Boston admissions for the remaining years in our series is likewise similar to that of the other hospitals. [Here a footnote account is given of a technical statistical analysis which showed that the differences between the age distributions are easily explainable by random or "chance" variation.] For the McLean Hospital we were not able to secure the age distribution of total admissions until 1876. This age distribution was virtually identical with that of the rest of the state and hence we have used the total age distribution for McLean in the earlier years as well.

What we required, of course, was the age distribution of *first* admissions rather than total admissions. Here, however, we were able to secure only scattered evidence primarily supplied by Dr. Earle in his Northampton reports. They provided, for several years, the age distribution of first admissions. This differed so little from the age distribution of total admissions that the use of the total admissions age distribution gave us a fully satisfactory basis for the calculation of age-specific first admission rates.* It should also be pointed out that in the period we are considering 65 to 75 percent of total admissions were first admissions. Consequently the age distribution of total admissions is in any case considerably weighted by first admissions. The final piece of evidence bearing on the use of the total admission age distribution for first admissions is provided by direct data on the age distribution of first admissions for 1880–85. This period in Table [55] is based on the exact reporting of

---

* It might be supposed that the average age of readmissions would normally be higher than the average age of first admissions. This is not, in fact, the case. Thus, during the contemporary period, both Illinois and Massachusetts show a *lower* age for readmissions than for first admissions. This is because patients with the psychoses of the senium [that is, old age] are much less likely to be discharged and hence to be readmitted than are patients who enter a hospital at an earlier age. In the 19th century, readmissions do not show a younger age of admission than first admissions because in this period (as we shall show shortly) admissions in the older age groups form a very much lower proportion of admissions than is the case today. [Footnote in original.]

TABLE 55

Age-Specific First-Admission Rates for Major Psychoses, Massachusetts, by 5-Year Periods, 1840–84 and 1885[a]

Age	1840–1844	1845–1849	1850–1854	1855–1859	1860–1864	1865–1869	1870–1874	1875–1879	1880–1884	1885
10–19	12.2	13.4	14.4	16.4	15.4	14.5	15.8	19.1	17.2	18.6
20–29	50.1	51.8	52.5	62.6	59.0	62.7	70.0	74.0	76.5	84.6
30–39	71.7	73.8	69.8	82.0	78.0	75.3	101.0	104.1	99.5	109.2
40–49	80.5	80.6	83.6	92.3	85.0	71.5	97.0	99.4	99.7	109.0
50–59	77.5	85.5	61.7	63.5	62.9	72.7	77.4	78.9	83.4	90.0
60–	50.1	59.9	44.5	48.0	68.0	60.5	66.8	68.0	80.2	67.8
Total	39.4	41.1	39.0	44.6	43.1	43.2	51.9	56.3	57.8	62.2

[a] *Ibid.*, p. 49. Detailed explanatory notes in the original are omitted here. Notice that the figures in the row labeled "Total" are rates for all ages combined from 10 up. These are not simple averages of the rates for separate ages, but weighted averages.

the age distribution of first admissions. We found that the assumptions we had used in the earlier years gave us an extremely striking continuity with the age distribution in this last period where direct evidence is available. Our method of estimation is therefore such that had we applied it to the period 1880–85 we would have come out with an age distribution that is virtually identical with that provided by the official reports.[9]

(d) Another question is the extent to which the figures on numbers admitted to hospitals reflect out-of-state admissions. It was possible to adjust quite accurately for this,[10] but we will not discuss it.

(e) A rate of incidence is a special kind of fraction. Therefore it is necessary to know not only the numerator (that is, the number of first admissions) but also the denominator (the total number of people in the relevant age group). The discussion up to this point has referred only to the numerator. The sizes of the total population and of the various age groups were obtained from Federal and state censuses.

> Although population enumeration was probably less accurate in the 19th century than today, the difference in the amount of error can hardly be such as to affect, to any appreciable degree, comparisons between the earlier and contemporary period.[11]

Analysis. One of the first findings was that the age pattern of first admissions was considerably different in the 19th century than today. That is, even if the average number of first admissions per 100,000 people in the total population had been the same at both times, and the age composition of the populations had been the same, the rates for specific age groups would have differed. In the 19th century, more of the first admissions occurred in the middle age groups, especially in the years 20 through 49, and fewer in the under-20 and over-60 age groups.

Next the authors present the following two tables:

[9] Goldhammer and Marshall, *op. cit.,* pp. 32–34.
[10] *Ibid.,* p. 34.
[11] *Ibid.,* pp. 34–35.

TABLE 57

AGE-SPECIFIC FIRST-ADMISSION RATES FOR MAJOR PSYCHOSES,
BY SEX, MASSACHUSETTS, 1880–84 AND 1885[12]

Age	1880–1884		1885	
	Male	Female	Male	Female
10–19	19.3	14.7	22.0	15.0
20–29	87.9	66.8	96.4	75.0
30–39	103.6	95.6	111.0	107.9
40–49	104.7	95.2	110.0	108.1
50–59	88.5	78.5	102.9	78.8
60–	84.8	74.5	70.4	65.5

[12] *Ibid.*, p. 50.

Many readers will be reading these lines after only a cursory glance at the tables. Actually, you can easily learn to read tables accurately and quickly, once you overcome this tendency to skim over or skip them entirely. It is unwise to be dependent upon someone else's interpretation of tables, just as it is unfortunate to be completely dependent upon an interpreter in dealing with a foreigner; and fortunately it is easier to learn to read tables than to learn a foreign language. There is danger that the main facts of the table may be obscured, if for no other reason than an author's desire to use variety in his wording when he is putting the facts into prose. It is not uncommon, either, for an author to misinterpret his own tables, or to overlook important matters shown by them. Moreover, tables can usually be read more quickly than verbal descriptions of them. The only explanation that should be needed for an interpretation of Tables 56 and 57 is that the rates are given on a base of 100,000 people. If, for example, there were 200 first admissions in a group of 200,000 people, the rate per 100,000 would be 100, that is, $(200/200,000) \times 100,000 = 100$. Note that the second table presents rates that are specific for both age and sex. The method by which this information on the sex distribution was obtained is discussed fully in the original report.[13]

[13] *Ibid.*, p. 48.

Now we let the authors resume the story:

We wish to test the hypothesis that in the central age groups the incidence of the major mental disorders has not increased over the last two to three generations. We bring to bear on this problem first admission rates for Massachusetts in the 19th century, and the question now arises: What rates from the contemporary period should be compared with them? The most immediate comparison that suggests itself is, of course, with the contemporary Massachusetts age-specific rates. This, however, is not necessarily the most desirable choice of comparative data. Hospital admission rates are a function (a) of the actual incidence of mental disease and (b) of factors that influence the proportion and type of mentally ill persons who are hospitalized. Our comparisons should, therefore, strive to ensure as much comparability as possible with respect to these latter factors, and where strict comparability cannot be attained, we must at least take them into account in our interpretations of the 19th century and contemporary rates.

The more important extraneous factors that need to be considered in testing the hypothesis are (a) level of hospital facilities relative to demand as measured, for example, by marked differences in the ratio of admissible patients who are rejected for lack of accommodations to the total number of admissions, or as measured by the sudden rise in admission rates resulting from the opening of new hospitals in the area most immediately accessible to them; (b) accessibility to the institutions as defined, for example, by the relation of admission rate to distance from a hospital (where other factors have been held constant); (c) motivation to use facilities for a given level of facilities available and accessible; (d) range or type of patients admitted, in terms of diagnostic class, degree of severity of the mental illness required to secure admission, and (partly related to this) whether admissions are for relatively long periods or just for a few days to permit observation or temporary care; (e) composition of the population with respect to factors (other than age) that influence admission rates both in terms of their relation to the foregoing factors and to the true incidence of mental disease, e.g., proportion of foreign-born and urban dwellers in the population. The large foreign-born (especially Irish) immigration of the mid-19th century renders it imperative to ensure that our 19th century rates, relative to those of today, are not rendered incomparable by differing proportions of the foreign-born population and different relations of foreign-born and native-born rates. . . .

Since the factors that influence the choice of a standard of comparison from the contemporary period were not constant through the period 1840–85, it follows that the rates chosen from the contemporary period for comparison with those of 1885 are not necessarily the appropriate ones to use in a comparison with 1860 or 1840. Consequently, in what follows we provide a variety of contemporary rates with which we compare the rates of different parts of our 19th century series; and in each case we indicate why these particular contemporary rates have been chosen for comparative purposes. The attempt to choose contemporary rates that provide the greatest constancy of the conditions (a) to (e) discussed above, must, to a considerable extent, be impressionistic. Sufficiently exact data on the factors involved, and on the weighting to be assigned to each, to permit the construction of a single quantitative measure, are not available. We have, however, in all cases chosen contemporary rates that we believe provide a severe test of the hypothesis under study—that is, we have chosen contemporary rates in which the operation of factors (a) to (e) are on the whole prejudicial to the hypothesis.[14]

In short, rates of first admissions to mental hospitals, even when given for specific age and sex groups, are not sufficient for making comparisons of the true incidence of mental disease if the *other factors* enumerated above are not comparable. The method pursued in the Goldhamer-Marshall study was dictated by the incompleteness of the evidence available as to the effect of the other factors on mental illness or on hospitalization for mental illness. The authors made a series of comparisons in which selected 19th-century rates were compared with selected 20th-century rates, the selections being made in such a way that the rates are affected in much the same way by the extraneous factors. We shall summarize the main comparisons actually made.[15]

(a) *Comparison of 1885 and Contemporary Massachusetts Rates.* In the first comparison the other factors were not really comparable, but were rather strongly "loaded" against the hypothesis the authors were establishing. Present-day Massa-

[14] *Ibid.*, pp. 50–52. Footnote omitted.
[15] *Ibid.*, pp. 53–76. The italicized headings are direct quotations.

chusetts first admission rates are higher than either the national average or the average for the New England States; a large part of contemporary admissions is for observation and temporary care; the percentage of urbanization is higher than in the 1880's; a larger proportion of current admissions are for nonpsychotic conditions. All these factors might be expected to exaggerate the magnitude of mental illness today by comparison with the 19th century. As might be expected, it turns out that the 19th-century rates are lower for most age groups, but, surprisingly enough, the rates for women in the 19th century were as high for ages 30–49 as they are today.

(b) *Comparison of 1885 Rates with Contemporary Massachusetts Rates for Admissions with Mental Disorder.* By using only those admitted "with mental disorders" in the calculation of current rates, it is possible to eliminate some of the artificial excess of current rates and hence make a fairer comparison. When this is done, ". . . The 1885 female rates exceed the contemporary rates for the entire age range 20–50. The 1885 male rates show substantial agreement with the contemporary rates for ages up to 40; the contemporary rate for the age group 40–50 is 13 percent in excess of the corresponding rate for 1885."[16]

(c) *Comparison of 1885 Massachusetts Rates with Contemporary Massachusetts Rates for Court and Voluntary Admissions.*

In this comparison we exclude the observation and temporary care first admissions, but include all regular admissions to public and private hospitals both with and without mental disorder [that is, in the contemporary figures]. . . . This comparison . . . reveals that the male 1885 rate for the combined age group 20–40 slightly exceeds that of the contemporary period and that the 1885 female rates for ages 20–50 exceed the corresponding 1930 figures. . . .[17]

(d) *Comparison of 1885 Massachusetts Rates with Contemporary Rates for Northeastern United States.*

[16] *Ibid.*, pp. 56–57.
[17] *Ibid.*, p. 58.

The comparison of Massachusetts late 19th century rates with those for contemporary Massachusetts imposes a quite severe test of our hypothesis. Nonetheless, for the central age groups, the hypothesis has stood up to the test applied. A further comparison that suggests itself is to juxtapose our late 19th century rates for Massachusetts with first admission rates for the combined Northeastern states (New England and Middle Atlantic states). The two preceding comparisons have provided somewhat greater comparability with respect to the classes of patients received in the two periods. In terms of comparability of level of facilities available, a better comparison can probably be achieved by using rates for a larger area in which the facilities may be presumed to deviate less from those of our late 19th century period. It would be desirable to provide a comparison which simultaneously attempts to equate level of facilities and class of patients, but the contemporary data do not permit this very readily. Our inability to make such a comparison means, of course, an increase in the severity of the test to which our hypothesis is subjected. In choosing the Northeastern states for our next comparison, we have by no means selected a low-rate area. These states have a first admission rate that is 20 percent above the average for the country as a whole. They are, taken together, highly urbanized and industrialized states and the great proportion of their population and admissions come from states that have well-developed mental hospital systems. Further, in making this comparison we include all admissions, both with and without mental disorder, to state, county, and city mental hospitals, Veterans Administration hospitals, and private mental hospitals. . . . Here again we find that the female rates for 1885 show complete parity with those of 1940 for the age groups 20–50. The male rates for 1885 show complete parity for the age groups 20–40; in the age group 40–50 the contemporary rate exceeds the 1885 rate by 17 percent.[18]

(e) *Comparison of Massachusetts 1855–59 Rates with United States 1940 Rates for First Admissions with Psychosis.* After a careful analysis the authors concluded:

> . . . the various conditions inhibiting admissions to mental hospitals were at least no less in Massachusetts of 1855–59 than they are currently in the United States as a whole. Probably

[18] *Ibid.*, pp. 59–61. Footnotes omitted.

there is no state in the Union today in which the restrictions on admission to institutions for the mentally ill approach those that existed in Massachusetts in 1855–59.[19]

The conclusion of this comparison was similar to the earlier ones.

(*f*) *Comparison of Massachusetts 1840–44 Rates with the Contemporary Period.* In the early 1840's, the Massachusetts rates of first admission undoubtedly understated greatly the true rates of mental illness because facilities were so extremely limited. "Some conception of the limited facilities available at this time is perhaps conveyed by the fact that during the first six years of its operation as the first state mental hospital, Worcester received additions of four wings which were no sooner completed than they were immediately 'filled to the overflowing.' "[20] The authors decide not to make precise tabular comparisons in this case, but by an "impressionistic" comparison with selected states which had low admission rates in 1940, they conclude that for women at least, "Given the restrictions on admissions in this earliest period it is quite impossible to suppose that this difference reflects a real increase in the incidence of mental disorders among women in these age groups in the intervening 100 years."[21]

(*g*) *Comparison of Suffolk County, Massachusetts, 19th Century Rates with Contemporary Rates for New York City.* Next, the authors turned to a slightly different kind of comparison:

So far we have dealt in our analysis with Massachusetts as a single unit. There are, however, several reasons why a special analysis of the Boston area recommends itself. In the first place, some interest is attached to the question whether . . . the disparity usually found today between rates for large metropolitan centers and for smaller towns and country areas existed in the earlier period. Secondly, and this is more important for our present purposes, we may presume that comparisons of 19th century and contemporary metropolitan rates provide a somewhat greater constancy of conditions than is feasible when

[19] *Ibid.*, p. 64. Footnote omitted.
[20] *Ibid.*, pp. 66–67.
[21] *Ibid.*, p. 68.

state rates as a whole are compared. Large urban centers probably have social characteristics that are more continuous over time. Perhaps more important is the fact that Boston residents, throughout our entire period, had two hospitals locally available . . . [these hospitals] did counteract to some extent the operation of the "law of distance."[22]

. . . As we are principally interested in the central age groups 20–50, our rates for these ages may . . . be taken as quite conservative estimates. We emphasize this because the rates we are about to present may astonish the reader and we wish to assure him that he is not dealing with inflated rates in the central age groups.

The only large urban center for which we were able to find contemporary first admission age-specific rates, including admissions to both state and private institutions, is New York City. . . .

. . . in 1840–44 Suffolk County had higher rates than contemporary New York City in the age group 40–60 and, by the mid-19th century period, almost uniformly higher rates except in the oldest age group. The reader may at this point feel that we have proved too much. We must confess that when these results became evident we ourselves felt intimidated by them. However a thorough reexamination of our data and our procedures has convinced us that these findings must stand.[23]

The final comparison, which we shall not discuss in detail, shows that the surprisingly high 19th-century rates are not unique to Massachusetts, although the best and most complete data are available for that state.

The remainder of the study may be described briefly with much less attention to the detailed evidence. The comparisons described above were concerned chiefly with the middle age groups, especially the ages 20–49. By contrast, the rates for older people are much higher today than in the 19th century. Yet the authors show that even this differential may possibly be due to "other factors" rather than to a "real" increase. They first point out that the preponderance of admissions of old people is

[22] By the "law of distance" the authors are referring to the tendency for the rate of hospitalization for mental disease to be greater for people who live closer to mental hospitals.
[23] Goldhamer and Marshall, *op. cit.*, pp. 68–73. Footnotes omitted.

due to senile and arteriosclerotic psychoses, "diseases of the senium." Therefore,

> Three major possibilities present themselves: (1) that there has been a true increase in the incidence of the arteriosclerotic psychoses and that consequently the admission rates for the older age groups have risen as a result of this; (2) that there has not been a true increase in the incidence of such mental diseases, but that the tendency to hospitalize such cases has increased; (3) that the increase in rates is a result of a combination of the two foregoing factors. . . .[24]

Medical research suggests that it is possible that a part of the increase in the senile psychoses is "real," but the evidence is not very conclusive. But there is fairly good evidence, statistical and impressionistic, that a much higher proportion of old people with psychoses are hospitalized today than in the 19th century. Hence the authors conclude:

> While not excluding the very real possibility that part of the increase in admissions in the oldest age groups is due to a true increase in arteriosclerosis, the foregoing considerations strongly suggest that a major share of the increase in the age-specific rates for arterosclerosis is due to the different hospitalization patterns for the older age groups in the 19th and 20th century periods. . . .[25]

The discrepancy between 19th- and 20th-century rates for those under 20 also turns out to be illusory. Contemporary admissions for this group include a larger proportion admitted for mental deficiency, as opposed to psychosis, than was true in the 19th century. "We conclude, therefore, that there is no evidence of an increase during the last century in the incidence of psychoses among persons under the age of 20, and that consequently the findings for ages 20–50 can now be stated to be true of all ages under 50."[26]

The authors also analyze the effect of the foreign-born popu-

[24] *Ibid.*, p. 77.
[25] *Ibid.*, p. 81.
[26] *Ibid.*, p. 83.

lation and conclude that this factor does not affect the conclusions already reached.[27]

We have frequently quoted the procedures used in making comparisons in order to convey without oversimplification the method used by the authors. Even so, we have omitted much of the careful discussion of sources of data and the footnote documentation for assertions made in the text. Such thorough attention to detail is certainly onerous and at times may seem overly pedantic. Yet the temptation to overlook details and to fill in with unsubstantiated assertions can easily lead to erroneous conclusions.

Conclusions

We have now described the analysis of the evidence. The main findings are summarized by the authors in five short paragraphs:

1. When appropriate comparisons are made which equate the class of patients received and the conditions affecting hospitalization of the mentally ill, age-specific first admission rates for ages under 50 are revealed to be just as high during the last half of the 19th century as they are today.

2. There has been a very marked increase in the age-specific admission rates in the older age groups. The greater part of this increase seems almost certainly to be due to an increased tendency to hospitalize persons suffering from the mental diseases of the senium. However, there is a possibility that some of the increase may be due to an actual increase in the incidence of arteriosclerosis.

3. The 19th and 20th century distributions of age-specific rates, that is, the distributions of admissions by age independent of changes in the age structure of the population, are radically different. In the 19th century there was relatively a much higher concentration of admissions in the age group 20–50; and today there is relatively a high concentration in the ages over 50 and more particularly over 60. This, of course, in no way affects the results summarized in paragraph (1) above.

4. Nineteenth century admissions to mental hospitals contain a larger proportion of psychotic cases and of severe derangement than do contemporary admissions. This is in part due to the more limited facilities of that period which tended

[27] *Ibid.*, pp. 83–89.

to restrict admissions to the severer cases, and to the different distribution of age-specific rates.

5. Male and female age-specific rates show a greater degree of equality in the 19th century than today. This is largely due to the differences discussed in paragraphs (3) and (4) above.[28]

Each of these conclusions is supported by the statistical evidence obtained. It is never safe to assume that conclusions are supported by evidence unless one actually examines carefully the evidence adduced. An author may state some evidence and then a conclusion without showing any connection.

As we indicated in discussing the purpose of the research, Goldhamer and Marshall are interested in much more than these findings. In their final section they discuss carefully the possible significance of their findings for problems other than those they attacked directly. In doing so they carefully distinguish between what they say on the basis of the evidence of this study and what they say on other grounds. We select a few sentences from this excellent discussion:

> In addition to random fluctuations, admission data do show short-term changes that coincide with marked social changes such as those incident to wars and depressions . . . our findings concerning the stability of secular trends for the psychoses [are] not intended in any way to minimize the importance of possible shorter-term fluctuations that have occurred in the past or may occur in the future.
>
> The necessary restriction of the findings to the psychoses, and more especially to those of a psychogenic or functional character, raises the question whether, nonetheless, the findings have any presumptive value for statements about long-term trends in the incidence of neuroses, psychoneuroses, and character disorders. . . . the implications of the present report, for views on long-term trends in neurosis rates, depend on the theoretical orientation to the neuroses that the reader favors.
>
> . . . A single study, such as the one reported here, can help to sharpen the formulation of alternatives, narrow the range of possible solutions to the theoretical problems at issue, and indicate promising directions for further research. Since the secular trend of admission rates has remained constant over the past 100 years, intensive research on short-term fluctuations is espe-

[28] *Ibid.*, pp. 91–92. In our description of the study, we have not brought out the evidence supporting the fifth conclusion.

cially indicated. This research will first need to determine whether these fluctuations represent true changes in incidence. If this is found to be so, it should then be possible to relate these rate changes to the specific alterations in life circumstances associated with the periods of changing rates. This would remove analysis from the level of the rather vague ascription of causation to broad social developments associated with the "growth of civilization" and lead to the analysis of the more concrete changes in social life that characterize the short-term periods under study. Only the combined and continuing research of laboratory, clinical, and social psychiatry can eventually enable us to discard those views that are inconsistent with observed fact. To this process the present report contributes the finding that, whatever may be the causal agents of the functional psychoses, they will almost certainly have to be sought for among those life conditions that are equally common to American life of a hundred years ago and today.[29]

This particular study has been described in such detail for several reasons. First, it is a competent, objective, and thorough investigation, and illustrates well the nature of such an inquiry. Second, the subject studied is important, but especially susceptible to erroneous impressions, hunches, and intuitions. Third, this investigation serves as an introduction to many of the basic problems encountered in statistical practice. Fourth, the statistical techniques used by Goldhamer and Marshall are elementary enough to be understood even before reading the rest of this book. Fifth, the study emphasizes the vital need for close integration of knowledge of subject matter with knowledge of statistical method, and for broad perspective on the general problem but meticulous treatment of the minutiae of the specific inquiry.

Vitamins and Endurance[30]

The Problem. By 1952 there was a good deal of evidence that extremely large doses of certain vitamins might enable animals

[29] *Ibid.*, pp. 92–97.
[30] This section is based upon Staff of Army Medical Nutrition Laboratory, *The Effect of Vitamin Supplementation on Physical Performance of Soldiers Residing in a Cold Environment* (Report No. 115 of Medical Nutrition Laboratory, Office of Surgeon General, United States Army, 15 September 1953. We have also drawn on unpublished statistical memoranda prepared during the investigation.

and possibly humans better to withstand severe physical and psychological stresses that exist under conditions of extreme cold. It had been reported, for example, that the ability of rats to continue swimming in very cold water (48° F.) was enhanced by vitamin supplementation of the diet. There was related, but less conclusive, evidence for human beings. On the basis of such evidence, the Canadian Army had decided upon vitamin supplementation for certain combat and survival rations. Supplementation on the scale needed (many times the normal requirements for the vitamins in question) was somewhat expensive, however, so the United States Army decided to conduct a special experiment involving simulation of battle conditions before supplementing its own combat rations. This experiment, which we shall describe in some detail, illustrates the care, persistence, and ingenuity needed to answer what at first appears to be a simple question, even though the investigators could specify what evidence they wanted, instead of being limited, as were Goldhamer and Marshall, to data that happened to have survived.

The objective of the experiment was "to determine the effect of supplementation with large amounts of ascorbic acid [vitamin C] and B-complex vitamins on the physical performance of soldiers engaging in a high-activity program in a cold environment, with and without caloric restriction." This objective is narrower than the original objective in two interesting and important ways. Originally, the intent was to ascertain the effect of supplementation on the physical performance of soldiers engaging in combat-type activity. The change to "a high-activity program" was necessary because of the near-impossibility of simulating combat conditions effectively; as we shall see later, this change was of considerable importance in interpreting the results of the experiment. Originally, also, the intent was to ascertain the separate effects of ascorbic acid (vitamin C) and B complex. The decision to narrow the objective to the study of the combined effect was dictated primarily by statistical considerations relating to the comparative smallness of the number of men available. We shall describe these considerations later.

The statistical problems of this investigation were anticipated in advance, when the study design could be molded to meet them. The scientific staff, though mostly M.D.'s and Ph.D.'s with specialization in physiology, were more conversant with statistical principles than are most research workers. Moreover, they worked closely with professional statisticians from the planning stage to the analysis and interpretation of the study.

Statistical Planning. Initially, it appeared that about 100 soldiers would be available, all volunteers and almost all from a Medical Corps establishment in Texas. Ideally, as we shall see in succeeding chapters, a random sample drawn from all combat soldiers in the Army would have given a better basis for generalizations beyond the group participating in the experiment. Such a random sample was, however, impossible for administrative reasons. The scientists had to make an extra-statistical decision, namely, that results for the population sampled would apply to the target population of interest—that is, that the sampling process actually used was satisfactory for studying the physiological response to vitamin supplementation.

All the men were to be housed in relatively insubstantial barracks, unheated during the night, in a cold and lonely spot in Wyoming, called Pole Mountain, at an elevation of 8,310 feet. Their clothing would be inadequate except when they were quite active. For most of ten weeks in January, February, and March 1953, they were to engage in strenuous outdoor activities: marches, forced marches, calisthenics, and sports. There were to be no leaves or passes. The diet was designed for monotony; the caloric total was ample at the start—4,100 calories—but a three-week period of short rations was scheduled for the end of the experiment, with only about 2,100 to 2,500 calories per day —about enough to maintain a stenographer's weight. It was anticipated that many of the men would collapse under the combination of strenuous activity and restricted diet, and the experiment would have to be terminated before the three-week period was over. Throughout the experiment there were to be periodic measurements of physical condition and performance, and also of psychological attitudes and aptitudes.

So much for the bleak regimen in store for the 100 volunteers. What about the statistical design? Your first reaction might be simply to give the vitamin supplements to everyone and see what happened. But when experiments are done that way, their findings are nearly valueless. The fatal defect is that no one knows how the men would have performed under these conditions in the absence of vitamin supplementation. The only way to find out what would happen without supplementation was to withhold the vitamins from some of the 100 soldiers. Then a comparison of performance could be made between those who did and those who did not receive the supplementation, the latter being called the "control group."

But which group of men should not receive the supplementation? The control and experimental groups should be such that, chance factors aside, both groups would react the same if treated the same. Then, if the supplemented group did better than the unsupplemented group even after allowance for chance factors, a decision could be made in favor of the vitamin supplementation. There is, basically, only one method of separating the men into the two groups so that the experimenter can draw valid conclusions about the effect of the supplementation: the separation should make proper use of *random sampling*. For example, the names of the 100 men could be put on slips of paper, the slips shuffled thoroughly, and the names for the control group selected by a blindfolded person. Random selection of the control group has two advantages. First, it protects against any bias of selection, conscious or unconscious, that might tend to make the control group systematically different from the other group. Second, only when the selection is essentially random is it possible to measure the influence of chance on the differences between the two groups, and so decide whether or not the actual difference exceeds that which would be expected from chance alone. These advantages will be explained in later chapters.

An important technical contribution of the statisticians to the design came about in the following way. The scientists suggested dividing the men randomly into four groups of 25 each. One group was to receive both vitamins C and B. The second was to receive vitamin C but not vitamin B, the third was to receive vitamin B but not vitamin C, and the fourth group was to receive neither vitamin. An alternative suggestion was to divide the 100 men at random into two groups of 50 each, one receiving *both* C and B, the other receiving *neither*. The statisticians recommended the second suggestion. With this two-group design, a more adequate evaluation could be made of the combined effect of vitamins C and B, though at the cost of not learning about the separate effects. With the four-group design, it was more likely that important *true* effects of vitamin supplementation would be obscured by chance factors, in which case a promising line of experimentation would be wrongly abandoned. If a significant effect for vitamins B and C together were detected by the experiment, further experiments to refine the findings by isolating individual effects would be inevitable. This reasoning was reinforced by the arbitrariness of the designation "vitamin B complex," which includes many distinct elements, each potentially as important as vitamin C. Hence even the original four-group experiment could not show which specific B-complex component was responsible for any effects of vitamin B complex. Moreover, in terms of the immediate military problem, the cost of multiple-ingredient vitamin capsules for front-line troops would not be much greater than that of capsules which contained only the effective ingredient or ingredients, since the cost of distribution from manufacturer to the Army and then to the troops represented the bulk of the total cost. Thus, if the combined B-complex and C supplementation proved effective, interim action could be taken, and later experiments could investigate more carefully the specific source of the benefits.

There was one major qualification in the recommendation of the statisticians that the vitamins be studied only in combination. This would be disastrous if, in truth, vitamin C and vitamin B complex each had beneficial effects, but the two together tended to cancel each other. The extent of this danger had to be evaluated by the scientists on the basis of their knowledge of the physiological effects of vitamins. They decided that the danger was remote, and adopted the two-group design.

An important question was whether 100 men were enough to make the experiment worth performing at all. The basic approach to this question was as follows: Suppose that the vitamins really have an effect which, if it could be detected despite inevitable chance variations, would be worth knowing about. What would be the probability of detecting such an effect in an experiment with 100 men? To answer the question, the statisticians needed to know (1) how big a difference would be "worth knowing about," and (2) how great the chance variation among men treated alike was apt to be. Both of these questions were studied in terms of one of the proposed measures of physical performance, the Army Physical Fitness Test, which consisted of five exercises: pull-ups, squat-jumps, push-ups, sit-ups, and squat-thrusts. This test had been used in the Army for several years with a standardized scoring system. It was known that an average improvement of about 20 points on this test might be expected during six weeks of the basic training period. If the same amount of improvement could be achieved merely by vitamin supplementation, supplementation would seem worthwhile. Next, records of past performance on the test were procured for a group of soldiers at an eastern camp. These records gave some idea about variation of scores on this test among individuals, and also about the variation of scores for the same individual on different occasions. This information made possible an assessment of the probabilities that such variation would obscure a true average effectiveness of supplementation of 20 units in an experiment based on 80, 100, or indeed any number of men that might be contemplated. It turned out that 100 men was about the smallest number for which an experiment would probably give valuable information.

Between the completion of the plans and the start of the experiment, the number of soldiers available was reduced to 87. This jeopardized the success of the experiment, but it was felt still to be worth doing. The previous decision to use two groups rather than four now seemed especially desirable.

In the planning period, several other issues were discussed by the statisticians and scientists. Some of them may not at first appear statistical, but all were relevant to the design of the experiment and to subsequent analysis of its results.

(1) It was essential to the success of the experiment that the soldiers themselves not learn who was receiving the supplements. Such knowledge might influence performance by its effect on the morale of the participants. It had been decided, therefore, to give capsules to everyone. The capsules for the control group had no nutritive value except for a trivial amount of vitamin C—just enough for normal requirements in the low-calory phase of the experiment. All capsules appeared identical in every respect but one: for ease of administration, capsules given to the supplemented group were colored orange while the others were colored brown. Thus, it would be known who was receiving the same treatment, although it would not be known what the treatments were. Even this knowledge could affect the experiment adversely. For example, one of the measures of physical performance was to be the ability to complete forced marches. Suppose that the first two men who fell out of a forced march turned out to be members of the group receiving capsules of the same color. Other men receiving capsules of this color might then suspect that they were not getting the superior treatment, and that they might soon have to drop out. Since, as the subsequent experiment confirmed, physical performance is very much influenced by attitude, falling out might become epidemic among those receiving capsules of the same color. Thus, what was really a matter of a few men being unable to continue would be exaggerated in the data because of psychological contagion. Moreover, all subsequent performance for the duration of the experiment might be strongly influenced by the memory of this one unhappy forced march. Unfortunately, it was not possible to correct this situation by making all capsules the same color. It was hoped that this defect in the design could be compensated by fostering strong inter-platoon competition in all the performance tests. Since each platoon contained men receiving capsules of both colors, men might identify themselves primarily with their own platoons rather than with their capsule colors.

(2) At the recommendation of the statisticians, all performance measurements were made once for each man before vitamin supplementation started. This permitted a more powerful analytical technique, based on the amount of improvement (or deterioration) of each man during the course of the experiment, rather than on his final performance alone. The importance of initial measurements of physical performance before supplementation was stressed by one of the statisticians in these words: ". . . failure to do so would be tantamount to removal of more than three quarters of the men from the experiment."

(3) The general strategy of the two-group design was modified to take into account the fact that the men were organized into four platoons, and that every effort would be made to foster inter-platoon competition. Instead of subdividing the entire group of men randomly into a supplemented and control group, a random subdivision was made within each platoon. This was as if four small experiments were performed instead of one big one, and there was reason to believe that the four small experiments combined would yield more reliable results than one big one.

(4) A still finer subdivision of the experiments by squads within platoons was considered and rejected.

Execution of the Experiment. Complex administrative problems arose in carrying out the experiment. A staff of 47 people —officers, enlisted men, and civilians—was needed, even though the subjects themselves handled the camp chores. The following jobs, among others, had to be done:

(1) Menus had to be devised to give the desired caloric values along with as much monotony as could be injected without causing excessive rejection of the food.

(2) The food had to be prepared with more than usual care in order that the theoretical caloric levels could actually be offered.

(3) All food not eaten by each subject had to be sorted and weighed in order to estimate his caloric intake, both in total and for protein, fat, and carbohydrate separately.

(4) The capsules had to be given to the right men, and it was necessary to be sure they were actually swallowed.

(5) Uniforms and barrack temperatures had to be adjusted to the weather.

(6) All activities and work details had to be scheduled properly.

(7) All performance tests had to be carefully supervised and recorded. For example, records had to be kept of the time and distance at which each man fell out on a forced march. Alertness was needed to notice such things as that fewer men fell out on forced marches if they were picked up and brought home in an open truck rather than a heated ambulance, and that still fewer gave up if they had to walk home anyway at their own pace. Total physical exhaustion was rarely encountered. "Experience with the forced march as a measure of performance, and specifically endurance, demonstrated that the usual cause of dropping out was loss of the will to proceed. It is not proper to call a man a quitter if he stops after marching 20 miles uphill into a fierce wind, yet in only rare instances did men apparently reach the limit of their capacity to march."

(8) Many special records had to be kept. For example, one Army enlisted man, trained as a meteorologist, kept detailed records of the weather.

(9) Twelve technicians in the laboratory section were needed to make the various physical and biochemical determinations, such as blood pressure, body weight, skinfold thickness, blood glucose, blood and urinary ascorbic acid, hemoglobin, and the like.

Analysis of the Findings. As the experiment drew to a close, attention was focused more closely on the details of the analysis. The general nature of the analysis had been determined before the experiment had even started, but there were many detailed questions to be answered. There were also innovations and improvisations in the experiment itself that had not been anticipated.

As data were collected in the field, rough analyses had been made by the supervising scientists, partly out of curiosity to see if the answer was going to be obvious. The most striking finding to emerge from these rough analyses was that the average physical performance for the entire group, supplemented and controls combined, had improved steadily throughout the experiment. In the last three weeks, when the 2,100–2,500 calorie diet had been expected to cause the experiment to terminate, the men not only carried on but continued to show improvement on the physical tests. When they departed on their "convalescent" furloughs, they were actually in better physical condition than at the start of the experiment. The unanticipated improvement of the men during the entire experiment, and especially that during the short-ration period, might have been attributed to the vitamins had there not been a control group which showed similar improvements. This outcome of the experiment thus underscores our earlier comments about the need for a control group.[31]

The answer to the basic question, then, was not obvious from the rough analysis. It would have been obvious only if the effect of the vitamin supplementation had been large and consistent. The actual differences were relatively small. Careful analysis was needed to decide whether the supplemented and control groups differed more than could reasonably be ascribed to chance.

As we have seen, there were many measures of physical status and performance. One of the most important was the Army Physical Fitness Test, described earlier. Initially, the combined fitness score—the sum of scores on the five components—was the focus for analysis. Before actual numerical work could begin, certain decisions about treatment of the data had to be made. The fitness test had been administered weekly during the experiment. A major problem arose because some of the subjects had missed an occasional test on account of injury or illness, or had participated when their physical conditions were below par for one of these reasons. When the latter occurred, a decision was typically made by the medical officers *before* the actual test whether or not the man's score would be included. However, six subjects presented more serious problems, and these were not finally resolved until the analysis was

---

[31] The need for controls is also illustrated by the experience of an elderly man who, having difficulty in hearing conversation, placed in his ear a plastic button with a cord long enough to run under his collar. Thereafter, he had no difficulty in hearing. People mistook the button and cord for a hearing aid, and talked louder. Had this man had a real hearing aid, he might have attributed all of the improvement in his hearing to the aid.

about to begin. To illustrate, we quote the description of two of these cases.

Test Subject No. 311: A thin, slight man of 22 developed an upper respiratory infection during the second week of capsule administration. . . . Soon thereafter, following vigorous physical exercise he developed a large hematoma in the right thigh. A pneumonitis ensued with fever, anorexia, vomiting, and 7½ pounds weight loss. He was at bed rest and light activity for approximately one month, a week of which was spent in the F. E. Warren AFB Station Hospital. During this time he missed four consecutive weeks of physical and metabolic tests. Following this illness his performance was generally poor and he continued to lose weight on the restricted caloric diet. It was decided to eliminate all of his data from the experiment. [This was the only subject for whom all data were discarded.]
Test Subject No. 432: This 30 year old platoon sergeant was granted emergency leave during the third week of the test . . . because of acute illness of several members of his family. He was absent from the test site for 10 days during which he administered nursing care to his family and continued to take capsules at the usual rate. No significant change of weight occurred during his absence, and tests of physical performance after his return showed no deterioration. It was decided to include all of the data collected from this man.

The final analysis, you will recall, was to be based essentially on improvement between the initial and final fitness test scores, and other performance measures. There were 44 men in the supplemented group and 40 in the control group for whom usable data were available for the first and last fitness tests. The results are shown in Table 78. The average score for the supplemented group was lower at the beginning and higher at the end; the average improvement was therefore greater for the supplemented group than for the controls.

On first glance, then, the supplementation appears effective. Actually, however, Table 78 shows only the over-all average for these two groups of 40 and 44 for the particular time period of the experiment. The table does not by itself tell whether these findings apply more generally. This question is what we had in mind earlier in our allusions to the effects of chance and the problem of allowing for those effects in interpreting the data.

It is possible to analyze the original data from which Table 78 was computed in order to reach a decision as to whether the greater improvement shown for the supplemented group is more than we would ordinarily expect by chance alone. The analysis used, though not the idea underlying it, is too technical for this book. The conclusion was that differences at least as great as the ones observed in Table 78 would arise purely by chance about 17 times in 100, *even if the supplementation had no effect.* The italicized clause, to use again the technical terminology first introduced in our discussion of the study of the incidence of major psychoses, expressed the null hypothesis. The evidence of the experiment is not strong enough to warrant discarding the null hypothesis, at least so far as this analysis was concerned.

TABLE 78

MEAN PHYSICAL PERFORMANCE SCORES OF SOLDIERS, INITIAL AND FINAL TESTS, VITAMIN-SUPPLEMENTED AND CONTROL GROUPS

Group	Initial Test	Final Test
Control	175.33	330.33
Supplemented	164.50	340.07

Several other analyses of the same type were made for other aspects of the fitness test data. For example, a separate analysis of each of the five component tests was made. In addition, analyses were devised which utilized not only the beginning and ending scores, but also the intermediate scores. None of these analyses provided convincing evidence against the null hypothesis.

The same analytical procedure was applied to several of the other physical and psychological tests. For some of these measures, the control, and for others the treated, group appeared slightly better. For the most part, the differences were readily ascribable to chance. On one type of test, however, the supplemented group appeared superior by a margin exceeding what would be expected by chance alone. The average drop in body temperature after periods of passive exposure to cold, both indoors and out, was less for the treated than for the control group. On the other hand, the loss of body weight during the experiment appeared significantly greater for the treated group.

Some of the measurements, such as performance on the forced marches, could not be analyzed by the approach just described because the data were qualitative (for example, a man did or did not fall out on a forced march) rather than quantitative (for example, scores on the fitness test). There was a variety of minor problems of analysis, but we shall report only the main conclusion: no convincing evidence in favor of supplementation.

This account may make the analysis sound tedious and complicated. It was. Moreover, many key questions arising during the analysis had to be handled by relatively crude statistical methods because more refined methods were not possible, given the then current state of statistical knowledge. There were a few interesting methodological by-products, statistical and medical, such as a better method for scoring the physical performance tests. Much was learned that would enable future experiments of this type to be more effectively conducted, and this was thoroughly discussed in the final report.

The most important criticism of the experiment was not a statistical one, by the problem of the meaning of *cold stress*. One crucial element of combat stress was missing: long, anxious, sleepless waiting in the cold. As the final report stated,

> The type and degree of cold stress should be precisely defined prior to the study and adhered to throughout . . . continued high energy activity is not compatible with body cooling despite the wearing of minimal uniforms. On the other hand, prolonged inactivity in the cold (simulating the fixed battlefield condition) is not compatible with high energy output. . . .

Our description of the experiment has necessarily neglected many important phases, but perhaps we have gone far enough to give you an appreciation of what underlay the brief statement of conclusions and recommendations, which we quote in full:

> Under the conditions of this experiment, supplementation of an adequate diet with large amounts of ascorbic acid and B-complex vitamins in men subjected to the stresses of high

physical activity, residence in a cold environment and, during the later part of the experiment, caloric deficit, did not result in significantly better physical performance than that of unsupplemented men.

Vitamin supplementation of the type used in this study resulted in a reduction in the fall in rectal temperature on exposure to cold.

A caloric deficit of 1,200 calories per day for 22 days did not lead to detectable impairment of physical performance.

The present study indicates that the current army minimal allowances of water soluble vitamins are capable of supporting good physical performance under the conditions of this study.

An ascorbic acid intake of about 60 mg per day (control group) resulted in whole blood ascorbic acid levels of 0.3 to 1.2 mg % with a mean value of 0.7 mg %.

RECOMMENDATIONS

1. That Army rations to be used in cold weather not be supplemented with ascorbic acid and B-complex vitamins. This recommendation is subject to change if further studies should reveal benefits not detected in the present study.

2. That further studies be made on the effect of vitamin supplementation on the physiological and pathological response of human subjects to cold exposure while at rest.

Artificial Rain-Making[32]

The next study to be considered in detail concerns a problem far removed from the evaluation of the effects of vitamin supplementation, but many of the statistical problems are surprisingly similar. The over-all objective of the research was "to obtain a more complete understanding of the fundamental physical processes which govern the formation of precipitation." The specific purpose of the part of the work reported here was to find out whether "seeding" cumulus clouds—that is, injecting appropriate materials, in this case water—causes the clouds to rain.

When the study began, the state of knowledge about cloud-

[32] This section is based on Roscoe R. Braham, Jr., Louis J. Battan, and Horace R. Byers, *Artificial Nucleation of Cumulus Clouds* (Chicago: Report No. 24, Cloud Physics Project, Department of Meteorology, University of Chicago, March 31, 1955). We have also drawn on unpublished statistical memoranda prepared in connection with this study.

seeding was not unlike that about vitamin supplementation when that study began, except that rain-making had attracted more public attention and claims of success were more numerous. Most of the evidence was sketchy and inconclusive, for a reason which may not be hard to guess if you think back to the vitamin study. Many clouds had been treated in many ways, but relatively little was known about what would have happened to them without treatment. It was almost as if the vitamin supplementation experiment had been performed without a control group and the inference drawn that supplementation was responsible for the physical performance observed.

In the present experiment, clouds were seeded by airplanes specially equipped with radar and photographic equipment, meteorological instruments, recording systems, and so on. Elaborate instrumentation was needed to obtain detailed information on all phases of cloud behavior. Moreover, the best way to detect the occurrence of rain was by the appearance of an "echo" on a radar screen. While radar equipment can tell whether or not a cloud produces precipitation, it does not tell how much precipitation is released or whether any of it reaches the ground. Thus, even if it could be shown that seeding initiated precipitation, it would not be known whether the seeding had simply altered the timing of precipitation that would have occurred later anyway, or whether it had increased the total amount of precipitation. Almost all attention in this experiment was focused on the simple question of initiation of precipitation, with a view to further study if initiation of precipitation were demonstrated.

The main statistical problem was to devise a method of deciding whether precipitation would have occurred in the absence of seeding. As in the experiment with vitamin supplementation, the need for a control group—unseeded clouds—was apparent to the scientists from the start. The precise way in which the control group was to be selected and the method by which the resulting data were to be analyzed were evolved during the course of consultation with several statisticians.

For reasons mentioned in the discussion of the vitamin experiment, it was essential that random selection be used in deciding which clouds were to be seeded and which were not. The main problem was to decide whether the proportion of seeded clouds producing rain differed from the proportion of unseeded clouds producing rain by a larger amount than could be ascribed reasonably to chance factors. Statistically, then, the basic problem in this experiment was similar to that of the vitamin study. One difference was that many of the measurements in the vitamin study were quantitative, for example, scores on the Army Physical Fitness Test, while in this study the basic measurements were qualitative, that is, rain or no rain.

In the vitamin experiment we saw that four small two-group experiments, one for each of four platoons, were preferable to one large two-group experiment. This was because the performance of soldiers within platoons was likely to be more homogeneous than the performance of all the soldiers in the experiment. Similarly, in the cloud-seeding study it seemed desirable to make each pair of clouds a separate small experiment. One cloud of each pair, selected at random, would be seeded. The two clouds in each pair, having been chosen at nearly the same time from the same part of the sky, would tend to be more like each other in respect to the probability of rain in the absence of seeding than would clouds selected at different times or different places. The analysis of an experiment performed in this way is simple, and later in this section we shall outline the idea briefly. Before coming to this, however, we must examine the remaining problems in the design stage.

The chief problem of execution in the paired-cloud design was to keep bias, conscious or unconscious, from entering into the selection of the cloud to be seeded in each pair. The following quotation from the report describes the procedure used:

> After a cloud had been selected for study, the senior scientist, acting as flight controller, instructed the meteorological-instrument engineer to release the treating reagent on the next pass. The cloud was treated or *not* treated depending upon further instructions available only to the meteorological-instrument engineer. The senior scientist who selected the clouds for study was physically isolated from other scientists and had no knowledge of which clouds were treated until after each

mission was completed. . . . Whether or not the cloud was treated, observations and measurements continued until it had dissipated, developed into a well-defined rainstorm, or lost its identity by merging with other clouds. . . .

Thus, the scientist who selected the clouds for study did not himself know which cloud in each pair had been treated. After the scientist ordered release of the reagent (water, in this experiment), the meteorological-instrument engineer opened a sealed envelope and read instructions which told him whether or not to execute the order. These envelopes had been prepared earlier by the statisticians, who assured random selection by using a method equivalent to tossing a fair coin. It was essential that the man selecting the second cloud in each pair not know whether or not the first cloud had been treated. Had he known, for example, that the first cloud had been treated, he might unconsciously have tended to pick a less (or more) juicy-looking cloud for the next pass. A systematic factor would then have worked for or against the treated clouds; this would have invalidated subsequent analysis based on the assumption that treated and untreated clouds differed only by chance.

The only conceivable way in which the controller might have known whether or not a cloud was treated was through the possible effect of the treatment on the behavior of the airplane. . . . When treating with water from the large valve, . . . the decrease in load on the airplane caused the plane to rise about 40 ft. in 14 seconds. . . . From the controller's position the change of altitude was detectable in clear still air, but totally undetectable in cloud or during airplane maneuvers. The reason for this lay in the natural turbulence found in the cloud and in the preoccupation of the controller with other duties.

As further substantiation of the fact that the effect of the release was not detectable during normal flight conditions, consider the outcome of flight 199, October 25, 1954. On this date, the crew obtained what they thought was a valid cloud pair, that is, the clouds met the eligibility requirements, echoes did not form in the interval between the inspection and treatment passes, and observation of the clouds continued for a satisfactory length of time. After landing, the ground crew started to refill the water tank and it was discovered that the valve had not opened. . . .

On the basis of all the evidence at hand, it is the unqualified judgment of the experimenters that the *controller* had no knowledge as to which cloud of any pair was treated until the information was revealed by the meteorological-instrument engineer after the test was completed.

Another problem of planning was to determine the number of cloud pairs needed in the experiment. Again, there is a close parallel with the vitamin experiment. The basic question, for any proposed number of cloud pairs, was the probability of detecting any given true effect of seeding in the midst of the effects of chance variation. This question could be handled less satisfactorily in the cloud-seeding experiment than in the vitamin experiment, because fewer performance data were available on unseeded clouds than on unsupplemented soldiers. Again, however, it appeared that an experiment of the scale permitted by available resources was large enough to give a reasonable prospect of informative results.

The most important results of the experiment are given in the following quotation:

> ... flight operations in the Caribbean area were carried out during two periods, October 1953 to February 1954 and October 1954 to November 1954. During the first period, 32 valid pairs of clouds were studied using the small water valve. The results are shown in Table [84].

TABLE 84

RESULTS OF TREATING CLOUDS WITH WATER USING SMALL VALVE
Total number of pairs—32

Untreated cloud of pair	Treated cloud of pair	
	Echo	No Echo
Echo	3	4
No Echo	3	22

Each unit in the table represents a pair of clouds, e.g., the number "4" in this table represents four pairs of clouds in each of which the untreated cloud produced an echo and the treated cloud did not. In the initial analysis, only the data in the lower left and upper right entries were considered. The numbers in the other diagonal represent those pairs of clouds of which both members behaved the same, and thus do not contribute to a test of the hypothesis that treatment makes no difference on the average.

From the fact that in three pairs the treated cloud developed an echo whereas the untreated did not, and in four pairs the reverse was true, it is obvious that the experiment does not support the efficacy of the treatment. A chance division of the 7 pairs could not be more even than 3 and 4.

The meteorologists suspected that the reason for the lack of effect might be the smallness of the amount of water being released. A larger valve was therefore installed, increasing the amount of water released. The large valve was used for the remainder of the operations. The results of precipitation initiation tests using the large valve are represented in Table 86.

The probabilities of obtaining the results in the lower left-upper right diagonal (or more unusual results) under the hypothesis of no treatment effects from the data in Table [86] (a), (b), and (c) are 0.11, 0.072 and 0.017 respectively. These probabilities were calculated under the assumption that large valve water treatment does not affect the average probability of precipitation initiation and are relevant if the alternative assumption under test is that water treatment increases the average probability of precipitation initiation. In designing this experiment, it had been decided that in problems of this type, 0.05 would be an acceptable level of significance. On this basis, it must be concluded from Table [86] (c) that the null hypothesis, i.e., that treatment had no effect, is false and that treated clouds had a higher probability of precipitation than untreated clouds. The probability of 0.017 given by the composite table must be viewed with some caution because it was calculated on the assumption that the total sample size was decided in advance, or at least decided on issues totally independent of what the initial results happened to be, whereas in fact this was not the case. It was decided to return to the Caribbean area after the data in Table [86] (a) were obtained. It is our opinion (and only an opinion since rigorous statistical techniques for this situation do not exist and since the rule for continuing experimentation was not made formal) that this will have little effect on the calculated probability. The experimental procedure and the types of clouds selected were the same during both seasons of operation.

TABLE 86

RESULTS OF TESTS FOR PRECIPITATION INITIATION IN TROPICAL CUMULUS CLOUDS TREATED WITH WATER USING LARGE VALVE

(a) January-February 1954; total number of pairs—15

Untreated cloud of pair	Treated cloud of pair	
	Echo	No Echo
Echo	3	1
No Echo	5	6

(b) October-November 1954; total number of pairs—31

Untreated cloud of pair	Treated cloud of pair	
	Echo	No Echo
Echo	2	5
No Echo	12	12

(c) All data; total number of pairs—46

Untreated cloud of pair	Treated cloud of pair	
	Echo	No Echo
Echo	5	6
No Echo	17	18

While the logic of the basic statistical analysis is clear in this quotation, it may be well to amplify it. In a group of paired tests, those pairs in which both clouds performed identically tell us nothing about whether treated or untreated clouds are more likely to rain. We are interested, therefore, only in pairs in which one cloud rained and the other did not. In the experiment summarized in Table 86(a), for example, there were six such pairs. In five of these pairs, the seeded cloud rained; in the sixth, the unseeded cloud rained. Now if seeding had no effect whatever, we would, except for chance, expect a 3–3 division rather than a 5–1 division. The situation is precisely analogous to an experiment you might do yourself with a coin. If you toss a fair coin six times, the most probable outcome is three heads and three tails. But chance alone will fairly often give you 4–2 or 2–4. How unusual, then, is the 5–1 result or one more extreme, that is, 6–0, on the basis of chance alone? If you have a fair coin and a little time, you can try this out and see how often you get either all heads or all heads but one in six tosses. If you do this enough times, and your coin is really fair, you will come very close to 0.11 or 11 percent, the figure given in the quotation.

Since 0.11 is a moderately small probability, we might suspect that 5 or more heads in 6 coin tosses was due not to chance alone, but to some inherent tendency for the coin to turn up heads more often than tails. Similarly, when the scientists in the experiment were confronted with the data of Table 86(a), they suspected that the treatment was actually having some effect. The suspicion was not very convincing, but it did lead to a renewal of experimentation in the fall of 1954. That time the division was 12–5, somewhat more convincing, since the probability is only 0.072 by chance alone. The combined evidence of the two experiments suggests quite strongly that seeding was effective.

We have quoted also some of the qualifications necessary in interpreting the final probability of 0.017 that a result at least this favorable to the cloud seeding procedure could arise by chance alone. This basic qualification is that the result of Table 86(a), the 5–1 split, was the first evidence of effectiveness given by the experiment. Earlier seeding efforts had given no hint whatever of success. Even though the experimental procedure had been modified prior to the 5–1 split, the experimenters could hardly put much confidence in this result alone. It is the evidence of Table 86(b) that is most convincing. Table 86(a) is not irrelevant to the final conclusion, but after all an event of probability 0.11 may well occur among a set of events, just as 5 or more heads in 6 coin tosses is fairly likely to occur if a number of such tosses are made.

A closely related point is that probably the experiment would not have been continued, so would have had no opportunity to produce Table 86(b), if just *one* cloud pair in Table 86(a) had been switched. The fact that the second phase of the experiment was run at all, then, partly reflects good luck in the first phase.

In this brief description, we have necessarily left out a great deal, even of the statistical problems. We have gone far enough, however, to illustrate the point made in the abstract in Chap. 1, namely, that the same statistical ideas are often applicable in problems that appear at first to be very different.

Conclusion

Problems that differ widely in subject matter may be essentially similar in the statistical methods they involve; yet statistical methods to be successful can never be applied in a mechanistic, formal way, but must be intimately interwoven into the whole fabric of the work. Effective use of statistics requires a continuous clear view of the important goals of an inquiry, and at the same time meticulous attention to innumerable details, many of which are of little interest in themselves except as they contribute to a successful pursuit of the major objectives.

Chapter 4

Misuses of Statistics

The Interpretation of Statistics

THE MOST important thing to know about the interpretation of statistical data is that they do have to be interpreted. They seldom if ever "speak for themselves." Statistical data in the raw simply furnish facts for someone to reason from. They can be extremely useful when carefully collected and critically interpreted. But unless handled with care, skill, and above all, objectivity, statistical data may seem to prove things which are not at all true.

"In earlier times," Stephen Leacock wrote, "they had no statistics, and so they had to fall back on lies. Hence the huge exaggerations of primitive literature—giants or miracles or wonders! They did it with lies and we do it with statistics; but it is all the same." Disraeli averred that there are three kinds of lies: lies, damned lies, and statistics. It is sometimes said that statistics are used the way a drunk uses a lamp post: for support, rather than for light. A famous statement about history has been paraphrased to say that the unsupported declaration "statistics prove" should be read "I choose to assert without evidence," or even "I choose to assert, contrary to the evidence." The view that statistical conclusions are usually wrong is often supplemented by the view that when they are not wrong they are self-evident and trivial: "A statistician is a person who draws a mathematically precise line from an unwarranted assumption to a foregone conclusion."

Misuses, unfortunately, are probably as common as valid uses of statistics. The ability to discriminate between a valid and an invalid use of statistics is more important for most people than knowing how themselves to make effective use of statistics. No one—administrator, executive, scientist, or responsible citizen in general—can afford to be misled by bad statistics; and everyone needs knowledge that can be gained only through the effective use of statistics.

Unfortunately, emphasis on misuses may give the mistaken impression that statistics are seldom or never reliable. Notice, however, that most misuses represent potentially good uses of statistics. "We share with Socrates the pious hope that men avoid mistakes once they are aware of them. But we are frivolous enough to suppose that men do this out of a spirit of pure contrariness, and hence are more affected by the sight of a horrible example than by a good precept."[1]

The examples which follow are divided into categories for purposes of discussion; the classifications are not to be taken very seriously, however, for many of the examples fall equally well into several categories.[2]

Misuses Due to Shifting Definitions

EXAMPLE 90A UNEMPLOYMENT IN DIFFERENT COUNTRIES

In some countries, no one is counted as unemployed unless he has been employed at some time in the past. In other countries, including the United States, anyone seeking work is counted as unemployed.

EXAMPLE 90B EMPLOYMENT, UNEMPLOYMENT, AND PARTIAL EMPLOYMENT

Late in 1949, Georgi Malenkov, then a member of the Soviet Politburo, asserted that there were 14 million unemployed in the United States, and that this showed that the United States was in a serious depression. (Actually, this figure would exceed by more than 1 million the U. S. Bureau of Labor Statistics'

[1] Ernst Wagemann, *Narrenspiegel der Statisik, Die Umrisse eines statistischen Weltbildes* (A Fool's Mirror of Statistics, the Outline of a Statistical View of the World) (3d ed.; Bern: A. Francke Ag. Verlag, 1950). We are indebted to Norma E. Kruskal for the translation from which the quotation is taken.
[2] Several of the examples which follow are taken from, or based on, those given by Jerome B. Cohen, "The Misuse of Statistics," *Journal of the American Statistical Association*, Vol. 33 (1938), pp. 657–674. Other excellent presentations of statistical fallacies are given by Darrell Huff, *How to Lie with Statistics* (New York: W. W. Norton and Company, Inc., 1954), and by Wagemann in the book referred to in the preceding footnote.

estimate of the average number unemployed in 1933, the highest in history, and it would fall only 2½ percent below the peak percentage—25, also in 1933.) Malenkov based his estimate on American data of good accuracy, but he defined "unemployed" to include all members of the labor force who worked less than full time. According to the definitions used by the U. S. Bureau of the Census, unemployment at this time was about 4 million. The basic difficulty here is that "employed" and "unemployed" do not cover all cases; they are extremes between which there is a wide range of possibilities. Furthermore, if there are many part-time workers, this may reflect either such scarcity of work that workers cannot find enough full-time employment, or such scarcity of labor that employers cannot find enough full-time workers. Full-time employment is itself hard to define, for hours of work that would now be considered full-time in America (for example, 35 to 40 hours per week) would be considered part-time in other countries (certainly in Mr. Malenkov's) or at other times.

EXAMPLE 91A   CAR REGISTRATIONS

Automobile registration figures are not an entirely satisfactory measure of the number of automobiles in the hands of the public, for three reasons. First, some states issue a new registration upon sale of a car, and some states transfer the old registration to the seller's new car, if any. Second, station wagons, sedan-type delivery cars, taxis, jeeps, and certain other types are classified as passenger cars in some states and not in others. Third, some cars are registered by dealers before they are sold to consumers. This last factor became important when the 1954 registration figures were scrutinized by two manufacturers each hoping to claim sales leadership. Enough unsold cars of both makes had been registered by dealers at the end of the year, to make it difficult to tell which make had led in sales.

EXAMPLE 91B   OVERHEAD COST

In studies of overhead and variable cost, confusion sometimes occurs between the economist's and the accountant's definitions of overhead cost. In economic analysis, overhead costs are those that do not change with the volume of output; accountants, however, sometimes allocate fixed costs among different years or different products in proportion to the volume of production.

EXAMPLE 92A   PERSONAL INCOME

Questionnaire studies of personal income, whether based on censuses or samples, usually suggest an aggregate personal income for the country as a whole which is at least 5 percent below the aggregate actually believed to be correct by economists. While the reasons for the understatement are complex, one basic cause is that people tend of think of income as wages and salaries only, rather than as their income from all sources.

EXAMPLE 92B   INDUSTRIAL CONCENTRATION

In measuring the extent of industrial concentration or "monopoly," the percentage of the total sales of an industry made by the four leading companies is often used. One difficulty with this measure, called the "concentration ratio," is that it may be affected greatly by the definition of "industry." If industries are defined relatively narrowly—for example, the "household electric toaster industry,"—the concentration ratio tends to be higher than if broader definitions are used—for example, the "household appliance industry," or the "electrical equipment industry."

EXAMPLE 92C   WAGE RATES AND WAGES EARNED

Although wages, in the sense of hourly wage rates, in a certain industry went up 10 percent, wages in the sense of average weekly earnings went down, due to a reduction in hours of work. Weekly earnings depend not only on average hourly earnings but on the number of hours worked. Wages may thus appear to have gone either up or down, or even both at once, if in defining wages no distinction is made between wage rates and wage payments.

EXAMPLE 92D   HOURLY WAGE RATES

In a labor dispute, the union presented figures showing that during a period when prices had risen, average hourly wage rates had fallen. The management presented figures showing that this average had risen. The management had averaged the straight-time rates of the individual workers (that is, rates for regular working hours rather than for overtime, holidays, nights, and so forth) and this average had increased. The union had gotten an hourly rate for each employee by dividing his earnings by the number of hours he worked. The decrease in the union's average represented a decrease in the proportion of the work that was done at overtime rates. Again, either conclusion may be correct, depending on which definition of hourly wages is appropriate. The union's definition will often be appropriate when wages are viewed as income of workers, the management's when they are viewed as costs of production. Which way they should be viewed depends upon the problem.

EXAMPLE 93A SEVERITY OF DISEASE

The stages of severity of a disease may be defined differently from one hospital to the next, and comparisons between hospitals are thereby made difficult.

EXAMPLE 93B DURATION OF LABOR

In a study designed to find factors related to the difficulty of labor in childbirth, "length of labor" was used as a measure of difficulty. One shortcoming of this definition is that the beginning of labor is sometimes not clearly defined.

EXAMPLE 93C LONDON VS. NEW YORK

Whether the city with the world's greatest population is New York or London depends on what areas are referred to by "New York" and "London." The City of London proper had a population in 1955 of only about 5,200, and New York County, or Manhattan, one of the five boroughs of New York City, had 1,910,000. The analogous political units, however, are the City of New York, with a population of 8,050,000 in 1955, and the County of London, 3,325,000 in 1955. Each of these is a municipality made up of boroughs, 29 in London and 5 in New York. A comparison often made, (though inaccurately) is that between Greater London and the City of New York—probably because of the coincidence that the City of New York, when it was formed by consolidation of New York, Brooklyn, and other areas in 1898, was referred to as "Greater New York." "Greater London," with a 1955 population of 8,315,000, is defined as the area within 15 miles of the center of the City of London. It has been estimated that the area within 15 miles of the center of New York has a population of 10,350,000. The "New York Standard Metropolitan Area," however, had a 1955 population of 13,630,000. (A Standard Metropolitan Area is defined by the U. S. Bureau of the Census as a county or group of counties containing at least one city of 50,000 or more, plus such contiguous counties as are metropolitan in character and integrated with the central city by certain specified criteria.) A metropolitan area defined for London on a basis similar to that used for New York would have a population of approximately 10,000,000.[3]

Misuses Due to Inaccurate Measurement Or Classification of Cases

When confronting statistical data, one useful question is, "How could they know?" Another is, "Who says so, and does he have a personal interest in the data being the way he reports them?" The answers to these questions do not settle anything about the quality of the data, for sometimes it is possible to find out about things for which the question, "How could they know?" suggests the answer, "They couldn't." (Would you have thought of the method described in Example 40C for finding out how many fish there are in a lake?) Moreover, some people are capable of great objectivity even when their own interests are involved. Nevertheless, the answers to these questions may properly stimulate skepticism.

EXAMPLE 94   CRIME RECORDS

Thomas F. Murphy, while Commissioner of Police in New York, issued a report that showed an increase of 34.8 percent in felonies in 1950 over 1949.

[3] This example was prepared for us by the Map Publication Editorial Department, Rand McNally and Company, Chicago, for which we are indebted to Duncan M. Fitchet.

This rise was one of the largest recorded in recent years. It was attributed by sources close to the commissioner's office ... [to] a recent overhauling of police customs in recording crimes and not ... [to] any appreciable increase in crime itself in the city. . . .

At police headquarters it was declared that the introduction of Mr. Murphy's system virtually eliminated the traditional "burying" of uninvestigated complaints or of unsolved crimes. . . .[4]

EXAMPLE 95A   INSPECTION ERRORS

In the inspection of manufactured products, sometimes every item is inspected and it is reported that all defective items have been eliminated. Actually, few inspections are completely accurate. Even with several inspectors each inspecting every item, some defective items are missed. More generally, measuring every one of a large group of things does not insure complete accuracy because of errors in the individual measurements.

EXAMPLE 95B   INFANT SEX RATIO

The sex ratio in live births is about 105 or 106 males to 100 females. Experts argue that this ratio may be a trifle high. Errors in reporting, editing, tabulating, and transcribing, though rare, tend to run predominantly in the direction of reporting girls as boys, or of omitting girls more often than boys. This effect is negligible for nearly all purposes, but the point is that errors do not necessarily—or even usually—cancel out.

EXAMPLE 95C   LANGUAGES OF THE WORLD

Consider the difficulty of obtaining information on the numbers speaking the various languages of the world. The following are the figures given by the 1950 editions of two well-known almanacs:[5]

[4] *New York Times*, June 7, 1951, p. 1.
[5] *New Yorker*, Vol. 26 (September 23, 1950), p. 80.

WHAT ALMANAC D'YA READ?

[From the World Almanac, 1950]

Tabulation of Those Who Speak the Chief Languages

Arabic	29,000,000
Chinese	488,573,000
Czech	7,500,000
Dutch	16,548,500
German	78,947,000
Hungarian	8,001,112
Italian	43,700,000
Japanese	97,700,000
Portuguese	48,800,000
Rumanian	19,400,000
Spanish	80,000,000
Swedish	6,266,000
Turkish	16,160,000

[From the Information Please Almanac, 1950]

Languages of the World

Arabic	58,000,000
Chinese	450,000,000
Czech	8,000,000
Dutch	10,000,000
German	100,000,000
Hungarian	13,000,000
Italian	50,000,000
Japanese	80,000,000
Portuguese	60,000,000
Rumanian	16,000,000
Spanish	110,000,000
Swedish	7,000,000
Turkish	18,000,000

EXAMPLE 96A INTERVIEWER EFFECT

The person who collects the data may consciously or unconsciously affect the response. For example, "When Negroes were asked if the army is unfair to Negroes, 35 percent said yes to Negro interviewers; only 11 percent said yes to white interviewers."[6]

EXAMPLE 96B DESTRUCTION OF PLANES

Accurate statistics about military operations are particularly difficult to obtain, even when great efforts are made. During the Battle of Britain in 1940, for example, the British estimated that the ratio of German to British air losses was 3 to 1. An American general was so impressed by the thoroughness of British methods that he believed the British claims of German losses were conservative. Yet a postwar check of German records showed that the correct ratio was 2 to 1.[7]

[6] Based on work done at the National Opinion Research Center, reported in the *University of Chicago Magazine*, April, 1952, p. 10.
[7] Winston S. Churchill, *Their Finest Hour* (Boston: Houghton Mifflin Company, 1949), pp. 337–339.

EXAMPLE 97A   DESTRUCTION BY PLANES

A similar example is given by the following quotation:

> Air attack by a single combat plane is a fleeting thing, and the results achieved do not always conform to first estimates. Air reports of destroyed vehicles, particularly armored vehicles, were always optimistic by far. This was not the fault of pilots. Each fighter-bomber airplane was equipped with a movie camera which automatically recorded the apparent results of every attack. The films were examined at bases and became the basis of "Air Claims," but we found that this method provided no accurate estimate of the damage actually inflicted. Exact appraisal could be made only after the area was captured by the ground troops.[8]

Misuses Due to Methods of Selecting Cases

EXAMPLE 97B   BRITISH TEXTILE UNEMPLOYMENT

During the early part of 1952, there was a slump in the textile industry of Lancashire. The extent of the decline was the subject of some controversy:

> Unemployment figures issued by the Ministry of Labor are misleading. They are based on counts of workers made on Mondays. But nearly all workers now employed only three days a week work on Monday and therefore are not included in the official short-time count. For example, the official count for mid-February shows 18,400 operatives on short time. However, the official "estimate," which is acknowledged to be correct by those in the know but that is not made public, gives the number of unemployed as 24,000 for the same period.[9]

Notice that whereas Example 90B involved matters of definition and interpretation—whether part-time workers are employed or unemployed, and whether they are symptomatic of prosperity or depression—this example hinges on the method of determining the number of part-time workers.

EXAMPLE 97C   CENSUS UNDERENUMERATION

In China, one census taken for military and taxation purposes showed a total population of only 28 million; but a few years

[8] Dwight D. Eisenhower, *Crusade in Europe* (Garden City, New York: Doubleday and Company, Inc., 1948), p. 324.
[9] *New York Times*, March 24, 1952.

later a census of the same territory for the purpose of famine relief showed 105 million. Such an increase could not possibly have actually occurred. People evade the census taker if taxes and military service are involved, but seek him out when it is a question of receiving aid. In general, census-taking is more difficult and the results less accurate than people commonly suppose. Even the United States census of 1950, for example, is reliably estimated to have understated the total population by 3.6 percent and to have erred by much larger percentages in its counts of some groups within the population.[10] In the capital city of one important Latin American country, the only census in recent years was abandoned after only two districts of the city had been canvassed, and the total population of the city is known only through intelligent "guesstimates."

EXAMPLE 98   MOVIE CENSORSHIP

The Chicago Police Department in 1952 prohibited the showing of the Italian film, "The Miracle." An interested organization reported the following investigation:

> In the past few months the Chicago Division has shown "The Miracle" at several private meetings. Of those filling out questionnaires after seeing the film, less than 1 percent felt it should be banned. "It thus seems," said Sanford I. Wolff, Chairman of the Chicago Division's Censorship Committee and Edward H. Meyerding, the Chicago ACLU's Executive Director, "that the five members of the Censorship Board do not represent the thinking of the majority of Chicago citizens."[11]

The statement quoted seems to be based on the assumption that those who saw the film at the private showings and filled out questionnaires represent the majority of citizens. Actually, it would be unwarranted to assume that the replies to the questionnaire represent the opinions of those who attended, to assume that those who attended were representative of those invited, or

[10] Ansley J. Coale, "The Population of the United States in 1950 Classified by Age, Sex, and Color—A Revision of Census Figures," *Journal of the American Statistical Association*, Vol. 50 (1955), pp. 16–54.
[11] *Civil Liberties* (published by the American Civil Liberties Union), December, 1952.

to assume that those invited were representative of the majority of citizens.

Many people find it hard to analyze separately the various elements in a complex and emotion-laden issue like this. For example, many who object to police censorship of moving pictures will resent our pointing out the statistical fallacies in this or any other attack on it, while many who approve the action will interpret our criticism of the attack as support of the action.

EXAMPLE 99A MENTAL DISEASE IN MEN AND WOMEN

The incidence of mental and nervous diseases appears to be higher among men than among women. A difficulty with the figures, however, is that men are more likely to be detected and institutionalized, since they are more likely to earn their livings in ways for which these disorders incapacitate them, and they are less likely to be supported by some other member of the family if unable to support themselves.

EXAMPLE 99B SCHOOL CHILDREN PER FAMILY

In a certain city, the average number of school-age children per family having school-age children was estimated by questioning a sample of children in schools. The figure obtained was much too high, because a greater proportion of large families than of small families was covered by the data. Consider two families, for example, one with a single school-age child and the other with six. The average number per family is seven divided by two, or $3\frac{1}{2}$. But if each of the seven children were asked the number of school-age children in his family, the total of the seven replies would be thirty-seven and the average $5\frac{2}{7}$. This example is discussed a little further in Chap. 6.

EXAMPLE 99C FAMILIES SELECTED THROUGH WAGE EARNERS

An error similar to that in the preceding example was made in estimating family earnings by sampling wage earners listed in employers' records. Those families with more than one wage earner had a greater probability of being included in the sample. Multiple-earner families tend to have higher incomes than single-earner families, not only because of the multiplicity of earners but because the heads of the families tend to be at the ages where earnings are highest, that is, at the ages where their children are old enough to earn money but have not yet left home.

EXAMPLE 100A   ERRORS OF EXECUTIVES

Consider the following statement by the head of a market research company:

> A "box score" which we have kept for a number of years shows that executives are right, or substantially right, only about 58 percent of the time in their decisions on questions of marketing policy and strategy. . . .[12]

The impression may be conveyed by this statement that executives arrive at the wrong conclusions 42 percent of the time when they solve their own problems. But the executives presumably bring to an outside consultant only the problems they consider beyond their own capacities. Thus, the market research company probably has a very biased sample from which to estimate the proportion of cases in which executives are right about marketing problems. In fact, the figures might even be interpreted as showing that the executives are wrong 58 percent of the time when they think they need outside advice on marketing problems! Moreover, there seems to be an assumption implicit in the quotation that the market research firm's answers are invariably correct, and this can hardly be quite true.

EXAMPLE 100B   AGES OF EXECUTIVES

This quotation is from a publication issued by a management consulting firm:

> The managements of a representative 65 companies are today, on the average, seven years older than were the managements

[12] Arthur C. Nielsen, "Evolution of Factual Techniques in Marketing Research," in Nugent Wedding (ed.), *Marketing Research and Business Management, University of Illinois Bulletin*, Vol. 49 (1952), pp. 52–53.

of these same companies 20 years ago. Here is what we found in a recent survey:

Average age of all officers	1929	1949
(excluding chairmen of boards)	47 years	54 years
Average age of presidents	53 years	59 years

In about 80 percent of these companies, those holding top management positions were older in 1949 than were their counterparts in 1929.

Taking the senior officers alone, i.e., presidents, vice-presidents, treasurers, controllers and secretaries, we found that they averaged 48 years of age in 1929 and 55 in 1949. More significantly the junior officers, who are normally regarded as replacements for the senior group, are not much younger than their superiors. They now average 52, and their advance in years since 1929 has followed the same upward trend as that of the senior officers in 65 companies studied.

With the average age of presidents today at 59 and that of all senior officers at 55, it is apparent that replacements will have to be made in the next five to ten years at a more rapid rate than has been the case in the past.[13]

The investigators are no doubt correct with respect to the particular companies studied (although note that the figures in the text and the table refer to different groups). These data do not, however, constitute evidence for or against the proposition that the average age of *all* executives has increased in the last twenty years. The proposition may be true, but these data do not show it. The fallacy is rooted in the fact that the same 65 companies were used in getting the average ages for 1929 and 1949. This means that the sample is limited to companies which have been in business for at least twenty years. Any generalizations made from the sample must therefore be restricted to companies which have been in business for twenty years. There is reason to presume, at least in the absence of evidence to the contrary, that the average age of executives in firms at least 20 years old is higher than in firms under 20 years old. Thus the method of selecting the sample is correlated with the very characteristic (age of executives) that is being studied. To study the change in average age one should select for 1929 a sample

[13] Booz, Allen and Hamilton, *Management Personnel: Is Your Company Building and Protecting its Most Valuable Asset?* (1949). Leaflet.

of corporations then in business, and then select for 1949 a different sample of the corporations then in business. Note that if a similar study had been made of ages of heads of families in 1929 and 1949, using the same procedures used in this study of ages of executives, similar but even more striking results would have been obtained.

EXAMPLE 102A LITERARY DIGEST

In 1936, the *Literary Digest*, a magazine that ceased publication in 1937, mailed 10,000,000 ballots on the presidential election. It received 2,300,000 returns, on the basis of which it confidently predicted that Alfred M. Landon would be elected. Actually, Franklin D. Roosevelt received 60 percent of the votes cast, one of the largest majorities in American presidential history. One difficulty was that those to whom the *Literary Digest's* ballots were mailed were not properly selected. They over-represented people with high incomes, and in the 1936 election there was a strong relation between income and party preference. In the preceding four elections, ballots obtained in the same way had correctly predicted the winners, but in those elections there was much less relation between income and party preference.

Misuses Due to Inappropriate Comparisons

This classification is closely related to misuses due to shifting definitions, due to shifting composition of groups, and due to misinterpretation of correlation and association.

EXAMPLE 102B POWER OUTPUT

On the same day, two New York papers published exactly contrary headlines. One stated that electric power output had gone up; the other that it had gone down. The first was comparing the power output of the current week to that of the preceding week; the second was comparing it to the corresponding week a year earlier.

EXAMPLE 102C EARNINGS AND RECEIPTS

In March, 1947, one New York paper headed an article "Douglas Aircraft Clears $2,000,000." Another paper carried the headline: "Douglas Aircraft Loses $2,000,000." The company had lost $2 million on its current operations, but had received $4 million refund on taxes paid in previous years.

EXAMPLE 103A NUMBERS OR PROPORTIONS OF ILLITERATES

The true statement that there are more illiterates in New York than in California requires further examination. There are many more people in New York, and on a percentage basis, we get the opposite conclusion. Whether to use absolute figures or percentages depends upon the particular purpose.

EXAMPLE 103B NUMBERS OR PROPORTIONS KILLED

During World War II, about 375 thousand people were killed in the United States by accidents and about 408 thousand were killed in the armed forces. From these figures, it has been argued that it was not much more dangerous to be overseas in the armed forces than to be at home. A more meaningful comparison, however, would consider rates, not numbers, of deaths, and would also consider the same age groups. This comparison would reflect adversely on the safety of the armed forces during the war—in fact, the armed forces death rate (about 12 per thousand men per year) was 15 to 20 times as high, per person per year, as the over-all civilian death rate from accidents (about 0.7 per thousand per year). Peacetime versions of the same fallacy are also common: "Homes are more dangerous than places of work, since more accidents occur at home." "Beds are the most dangerous things in the world, because more people die in bed than anywhere else." "Sick people are more likely to die when cared for in hospitals than when cared for at home."

EXAMPLE 103C PLEASANT AND UNPLEASANT WORDS

A psychologist found that a group of young children used "pleasant" words much more frequently than "unpleasant" words. From this finding it was concluded that children learn pleasant words more easily and rapidly than unpleasant words. It would be better to define "ease of learning" by the ratio of "pleasant" or "unpleasant" words actually learned to the total number of "pleasant" or "unpleasant" words which the children had equal opportunities to learn.

EXAMPLE 104A HEREDITY VS. ENVIRONMENT

Sometimes people try to make quantitative comparisons that cannot possibly be meaningful. One illustration is the heredity-versus-environment controversy. The following quotation shows how a logically meaningless comparison can give rise to a meaningful question:

> Hence it is really not legitimate to ask: What is the relative importance of heredity and environment? This question belongs in the scrap basket with the type of general conclusions in some of the studies quoted: "It appears that heredity is twice as important as environment in determining intelligence." The new approach would be: Given a stated environment, how much variation will heredity permit for such and such a characteristic (among so and so individuals)? Or, given a stated heredity, how much variation could a given range of environment introduce for such and such a character.[14]

EXAMPLE 104B DIVORCE RATES

After the 1930 census, it was stated on the basis of tabulations of one-half of the states, that the divorce rate had apparently fallen from 1920 to 1930. When all the results were tabulated, it was found that the divorce rate had not changed. The error was due to the fact that the first states reporting were the less populous, agricultural, lower-divorce-rate states. These states should have been compared with the same states in 1920, instead of with the whole country in 1920, though even then the result could have been interpreted only as applying to these states, not the whole country.

EXAMPLE 104C INCOMES AND PRICES

In the 1936 election, the Democrats claimed that employment and production had risen greatly while the cost of living had not gone up at all. Their bases for comparison were 1933 for employment and production and 1925–29 for cost of living.

---

[14] Gladys C. Schwesinger, *Heredity and Environment: Studies in the Genesis of Psychological Characteristics* (New York: Macmillan Company, 1933), p. 459.

The Republicans, on the other hand, claimed that the cost of living had gone up, but not employment. They compared cost of living with 1933 and employment with the 1925–29 average.

EXAMPLE 105A POSTWAR JAPANESE PRODUCTION

In 1949, an article in *Fortune* criticized the American regime in Japan for its handling of economic problems. It claimed that industry in Japan was stagnant by comparison with prewar production. The reply was made that Japan had made a greater improvement since 1946 than any other country in the world. The disagreement about the economic status of Japan turned chiefly on the base date for comparison.

EXAMPLE 105B RUSSIAN DOCTORS

"According to Yaroslavsky, the number of doctors in Russia had increased from 1,380 in 1897 to 12,000 in 1935."[15] In studying the effectiveness of the Soviet regime in expanding the number of doctors, it would be desirable to compare the increase between, say, 1897 and 1916, with that between 1917 and 1935, since the regime took over in 1917.

EXAMPLE 105C PRICES DURING AND AFTER CONTROL

The Office of Price Administration based its claims of effectiveness in holding down prices on the Bureau of Labor Statistics' Cost of Living Index (now called the Consumer Price Index); but after OPA was discontinued in July, 1946, some of its supporters showed alarming price rises on the basis of the same Bureau's Index of Spot Primary Market Prices of 22 Commodities. A wholesale price index generally fluctuates more than a consumer price index, and this specific index, based on daily quotations of 22 raw materials, fluctuates more than the Bureau of Labor Statistics' Wholesale Price Index, which is based on weekly and monthly average prices of 2,000 commodities (only 900 at the time the OPA ended) of all kinds. At this particular time, moreover, most of the rise in the spot prices index was due to one transaction (and that by the government) in one commodity, silk.

[15] Bernard Pares, *Russia* (New York: Mentor Books, 1949), p. 137.

EXAMPLE 106A PROPORTION OF CHINA LOST

In mid-1949 the following argument was adduced to support the position that the Chinese Nationalist government was not yet defeated in the war against the Communists on the Chinese mainland: "The Nationalists retain control of about half the area that is China. The Communists hold no more territory than the Japanese held at the height of the occupation." The mere fact that the Nationalists held about 50 percent of the territory of China tells nothing at all about the proportion they held of population, important cities, resources, or transportation facilities.

EXAMPLE 106B PROPORTION OF UNITED STATES VULNERABLE

In 1954, a statistically similar example appeared in a Chicago newspaper article, which attributed the following statement to a leading civil defense official:

> Even if Russian planes destroy the 70 largest industrial groups of cities, it will knock out only 3 percent of the nation's real estate. Ninety-seven percent would still be ready for business.[16]

Misuses Due to Shifting Composition of Groups

This category is closely related to inappropriate comparisons and to misinterpretation of correlation or association.

EXAMPLE 106C GROUP AVERAGE DOWN, EACH INDIVIDUAL UP (OR OUT)

A manufacturing plant found that the average monthly earnings of its employees had fallen 8 percent during a certain period. This might seem to "prove" that earnings had gone down. As a matter of fact, however, the earnings of every single employee were exactly 10 percent higher than at the beginning of the period. The reason the average earnings fell despite this increase was that many of the higher-paid employees were dropped at the time the increase was made, so that the new average included only the lower-paid workers.

[16] *Chicago Daily News*, September 28, 1954.

EXAMPLE 107A   OLD GRADS

The alumni of a certain class of a college had an average age of 87 one year and 85 the next year. The explanation is not that they were literally getting younger every year, but that the oldest members had died during the year.

EXAMPLE 107B   ARIZONA TUBERCULOSIS DEATH RATES

The death rate from tuberculosis is far higher in Arizona than in any other state. This might seem to indicate that Arizona has a bad climate for tuberculosis. Actually, it reflects the fact that its climate is considered beneficial by many people who have tuberculosis, who therefore go there. For causes of death such as heart disease, cancer, and cerebral hemorrhage (which are, in order, the three leading causes of death in the United States) Arizona has extremely low death rates—largely because it has a relatively young population.

EXAMPLE 107C   REGIONAL DIFFERENCES IN INCOME

Regional comparisons of income reveal a difference between average income in the North and in the South. An analysis of the North alone, however, reveals differences in average income between white and colored and between urban and rural areas. An analysis of the South alone reveals similar intraregional variation. Since the proportions of the population colored and white or urban and rural are not the same in the North and South, the average incomes of the two regions would be different even if the incomes of corresponding groups were the same in the North and the South. Part of the difference between the two regions is thus due to factors that operate within regions; this part simply reflects differences in racial composition and in urbanization, rather than differences in income of corresponding groups in the two regions.

Misuses Due to Misinterpretation of Association or Correction

This kind of misuse is really a special case of inappropriate comparisons. It exemplifies the familiar but often ignored fact that correlation or association does not necessarily indicate causation.

EXAMPLE 108A FEET AND HANDWRITING

In a study of schoolboys, an educator discovered a correlation between size of feet and quality of handwriting. The boys with larger feet were, on the average, older.

EXAMPLE 108B STORKS' NESTS

There is reported to be a positive correlation between the number of storks' nests and the number of births in northwestern Europe. Only the most romantic would contend that this indicates that the stork legend is true. A more prosaic interpretation is this: as population and hence the number of buildings increases, the number of places for storks to nest increases.

EXAMPLE 108C PROPAGANDA LEAFLETS

During the Italian campaign of World War II, it was found that there was a positive correlation between the number of propaganda leaflets dropped on the Germans and the amount of territory captured from them. While this is consistent with the hypothesis that the leaflets were effective, it is also consistent with other hypotheses, for example, that leaflets were dropped when major offensives were about to begin.

EXAMPLE 108D BUSINESS SCHOOL ALUMNI

A study of the alumni of a certain university's graduate school of business showed that students whose grades had been about average had higher incomes, on the average, than either the poor or the excellent students who graduated at the same time. Before drawing conclusions from this, it would be necessary to investigate the possibility that a higher proportion of the excellent than of the average have gone into teaching, where earnings are less than in business.

EXAMPLE 108E KENNY TREATMENT

Sister Elizabeth Kenny, originator of a method of treating poliomyelitis, declared yesterday that if a true knowledge of her method were available to the medical world, recoveries from the disease would be increased at least 10 percent.

The Australian asserted that when she arrived in this country in 1940, the percentage of recoveries was about 15 percent and that since, the recovery rate had risen to about 75 percent.

"This could be increased to 85 percent if the medical profession had the full knowledge of the Kenny treatment," she declared. "I'm not saying this . . . statistics already published prove it."[17]

From the newspaper article it is impossible to be sure what statistics Sister Kenny was talking about. It appears, however, that she was attributing much of the improvement after 1940 to her method of therapy. It is entirely possible that other things which happened after 1940 might account for some or all of the increase in recovery rate. For one thing, the diagnosis of polio improved so much that it became possible to detect many more mild cases. This improvement in diagnosis had the effect of increasing by equal amounts both the numerator and denominator of the ratio of recoveries to total cases. Hence, the recovery rate seemed to be increased.

Misuses Due to Disregard of Dispersion

EXAMPLE 109A CALIFORNIA WEATHER

In California, the weather is usually called unusual, as though the average were the usual value. Actually, substantial departures from the average, especially in winter rainfall, are typical of California. In some cases, it may actually be impossible for any single observation to be equal to the average, as when the average number of persons in a certain category is 2.38.

EXAMPLE 109B WADING IN THE TOMBIGBEE RIVER

Congressman John Jennings, Jr., of Tennessee, in the U. S. House of Representatives on June 6, 1946, stated that in dry weather the average depth of the Tombigbee River is only one foot. "In other words," he said, "you can wade up it from its

[17] *New York Times,* August 26, 1949, p. 12.

mouth to the spring branch in which it originates." While such a wading trip may be possible—we are assured by a native of the region that it is possible, although he has not himself made the trip—it does not follow from the statement about average depth. Something would have to be known about dispersion.

EXAMPLE 110A   MINIMUM SALARY SCALE

The President of an institution proposed to set $12,000 as the minimum annual salary for a certain class of employees. He asked an assistant to calculate the addition to the payroll that would result. The assistant found that there were 250 such employees and that their average salary was $11,000. He therefore reported that the cost would be 250 times $1,000, or $250,000 per year. Actually, the cost turned out to be $450,000. To see the point, suppose there were 50 employees at each of the following salaries: $7,000, $9,000, $11,000, $13,000, and $15,000. The average is $11,000. The increase in the payroll caused by each group would be $250,000, $150,000, $50,000, $0, and $0 respectively. The figure of $250,000 that the assistant calculated is what it would cost to raise the *average* to $12,000.

EXAMPLE 110B   SASKATCHEWAN WHEAT

One farmer [in Saskatchewan] has reported harvesting 104 bushels of wheat from a two-acre strip. Thirty-five bushels an acre is considered well above the average.[18]

Even if 35 bushels is a high average, it does not follow that 52 bushels per acre is unusually high for the best two-acre strip in a large area. It might even be unusually low. Averages based on large numbers vary much less than individual measurements, or than averages of very few measurements Especially when an observation is selected as being the most unusual, the average is almost useless as a good bench mark against which to judge its unusualness.

The fact that Herbert Hoover has lived more than a third

[18] *New York Times,* September 1, 1955, p. 1.

of a century after his inauguration as President (1929) does not by itself permit the inference that Presidents of the United States have unusual longevity. Similarly, the claim that Brand A gives "up to" 2½ times as much wear as the average of three leading competitors does not mean that Brand A gives more wear on the average.

Misuses Due to Technical Errors

The preceding statistical misuses, though frequent, are relatively obvious—at least after they have been pointed out. The errors are errors of common sense or of logic more than of statistics in any very technical sense. It is their frequency in statistical applications that justifies their emphasis here. There are, however, misuses that arise from more technical statistical deficiencies. Misuses of this kind will be discussed from time to time in later chapters. For the present, we give some illustrations that show the ever present danger of the most prosaic but most common technical error, a mistake in calculation.

EXAMPLE 111A   ERRORS IN COMPUTING STANDARD ERRORS

Ericksen's failure to find statistically significant differences between groups was due to erroneous computation of the standard errors of the differences between means. Instead of using the standard errors of the separate means, he used the standard deviations of the score distributions; hence, all of his 9 reported critical ratios tend to be quite low, 0.32 or less. With the use of the proper formula, some of the logically expected differences are statistically significant.....[19]

EXAMPLE 111B   ERRORS IN COMPUTING
AVERAGE PERCENTAGE

A firm manufacturing complicated electrical devices had found that it had to expect about 5 percent of the units made to be defective, but that the rate need never exceed 10 percent with good materials, machines adjusted properly, and skillful workmanship. One week, more than 10 percent defective units were

[19] Quinn McNemar, "Opinion-Attitude Methodology," *Psychological Bulletin*, Vol. III (1946), p. 304.

reported, so special care was given the next week's production. Nevertheless, 16.4 percent defective units were reported. One of the engineers called in to make an intensive survey of the production line looked at the inspector's records and found:

Day	Number Inspected	Number Defective	Percent Defective
Monday	70	0	0
Tuesday	68	2	3.0
Wednesday	68	3	4.4
Thursday	70	1	1.4
Friday	72	4	5.5
Saturday	32	1	3.1
Total	380	11	16.4

The correct percent defective was therefore, 11/380 or 2.9 percent. The inspector had used a wrong method of calculating—adding the daily percentages—and furthermore had added wrong and made two small errors in calculating the daily percentages!

EXAMPLE 112   ERRORS IN UNITS OF MEASUREMENT

During World War II, military and scientific people developing a promising new bombing device were disheartened when a statistician's calculations showed that the device would have virtually no chance of hitting its targets. Those responsible for the project were hastily gathered together from all parts of the country to consider this bombshell. Another statistician noticed that it would be physically impossible for a bomb to get as far from the target as the average error shown by the calculations. Hurried long-distance phoning and frantic all-night checking revealed that in the computations angular errors had been measured in degrees, but interpreted as if they were in radians (a radian is 57.3 degrees).

A frequent error of the same kind is to confuse two kinds of logarithms, "natural" and "common," the former being 2.3 times the latter.

Misuses Due to Misleading Forms of Statement

EXAMPLE 113A   CO-EDS MARRYING FACULTY

The statement, "One-third of the women students at Johns Hopkins University during its first year married faculty members," creates an impression unwarranted by the facts. There were only three women students. Similarly: "Thirty-three percent of the women married two percent of the men."

EXAMPLE 113B   CRAZY RADAR MECHANICS

The preceding example led a World War II veteran to tell us of an effective use he had made of the same form of statement. During the war, he was responsible for airborne radar in the Mediterranean Theater. He was able to obtain only seven radar mechanics for the Troop Carrier Command, which was authorized to have, and badly needed, forty to fifty. Repeated requests and complaints submitted through normal channels accomplished nothing. "One month I was informed," the veteran told us, "that one of the seven mechanics had suffered a mental breakdown precipitated by overwork. In my next monthly report, under the heading 'Troop Carrier Command—Personnel' all I wrote was 'Over fourteen percent of the radar mechanics went crazy last month due to overwork.' Almost immediately after the report reached Washington, thirty-five additional radar mechanics were sent us by high priority air."

EXAMPLE 113C   PALO ALTO SUMMER RAIN

At Palo Alto, California on July 25, 1946, nineteen times as much rain fell between 6 A.M. and noon as during all the preceding Julies since the weather station opened in 1910. That is, in six hours 19 times as much rain fell as in a 26,784-hour period, a rate during those six hours about 85,000 times "normal." Actually, this "deluge" consisted of only 0.19 inches; the only measurable rain in all the thirty-six previous Julies was on one occasion when 0.01 inches fell.

EXAMPLE 114A   GROWTH OF CHILDREN

> PALO ALTO, Calif., Aug. 30 (Science Service)—How tall a
> growing child will be when he is grown up is now being pre-
> dicted to within a quarter of an inch by scientists at the Leland
> Stanford University here. . . . [The scientists] report eight cases
> in which the adult heights came to within one-quarter of an inch
> of the heights predicted while the subjects were children. . . .[20]

The first sentence of this quotation leads the reader to believe
the implausible proposition that the scientists can predict *any*
child's height to within one-quarter of an inch. But the second
sentence suggests that they may be able to do no more than
anyone else, namely, be right occasionally—say 8 times in several hundred.

EXAMPLE 114B   PAJAMA SALES

> Pulling its drawstring tighter, the men's pajama industry dole-
> fully reported some raw facts last week: Men were buying only
> one-third of a pair of pajamas each a year. . . .[21]

The statistical finding, of course, is that an average of one-third
of a pair of pajamas per man was sold. While hardly anyone
would be misled by this example, which is attributable to jour-
nalistic flippancy, statements of this kind often convey the
impression that statisticians are chiefly concerned with quaint
curiosities, and warrant jibes like the following: "The average
statistician is married to 7/10 of a wife, who tries her level best to
drag him out of the house 2¼ nights a week with only 50 percent success . . .[etc., etc., etc.]."

EXAMPLE 114C   WORLD HEALTH ORGANIZATION

> Each American citizen contributes slightly more than 2 ct/yr
> toward support of the World Health Organization.[22]

This is a way of making a sum apparently in excess of $3 million
per year sound small. It would have sounded still smaller if it
had been expressed as a twenty-fifth of a cent per week or a
two-hundredth of a cent per day. Literally, of course, no individual American citizen contributes anything toward support of

[20] *New York Times*, August 31, 1949, p. 20.
[21] *Newsweek*, August 1, 1949, pp. 50–51.
[22] *Science*, Vol. 120 (1954), p. 955.

the World Health Organization. Whatever amount the American government contributes is derived from a number of sources, including taxes levied on citizens and others.

EXAMPLE 115A 1948 GALLUP POLL

In 1948 the American Institute of Public Opinion (usually called the Gallup poll) predicted confidently, on the basis of a series of polls culminating in one involving about 3,000 interviews with what it called a "scientifically" selected sample of voters, that Thomas E. Dewey would be elected president by a substantial margin. Other polling organizations made the same prediction, some even more confidently. Actually, Harry S. Truman was elected by a small margin. Many factors combined to produce this failure, a major one being statements and interpretations based as much on assumptions as on the data. A confident prediction in favor of either candidate was not justified by the data, but the prediction was based on a number of assumptions, including important ones concerning the voting of the substantial group who had not made up their minds when interviewed. In the preceding three elections, similar methods had successfully predicted the winner, though there had been large errors in the margins predicted, and this may have been conducive to carelessness in statements made in 1948. Incidentally, the upshot of the 1948 failure was that in 1952 Gallup presented his data without predicting a winner. The actual winner was Dwight D. Eisenhower, and his margin, 10.5 percent of the 62 million votes cast, was substantial enough—it was exceeded in only 7 of the 17 elections from 1888 to 1952—so that, to be useful, a method of forecasting presidential elections would have to be able to detect it.

EXAMPLE 115B OMISSIONS

Forecasters of business (and other phenomena) sometimes relate successes which, if they were the only forecasts made, would be strong evidence of prognosticating ability, but which in fact are selected from a long list of forecasts that, in the aggregate, are unimpressive despite occasional successes. In general, conclusions that would be correct if the figures cited were the only ones relevant, may be seriously qualified or even reversed if the complete data are taken into account.

The principle involved can be brought out by a simple example. Suppose you were told that a certain coin had been tossed 10 times and had showed heads each time. You would be fairly confident that the coin would continue to show more heads than tails, for the probability is a trifle less than 1 in 500 that a fair coin would show the same side on 10 tosses. But now suppose the fact was that 1,000 coins had each been tossed 10 times, and this one selected afterwards because it had shown heads all 10 times. If you had known about the other 999 coins, and the method of selecting the one about which you were told, you would not have felt much confidence that the coin would continue to show more heads than tails; for six times out of seven, on the average, sets of 1,000 fair coins tossed 10 times each, would have one coin that falls the same way all 10 times.

As the courts have recognized since time immemorial, it is not sufficient to draw conclusions from the truth and nothing but the truth; it must also be the whole truth.

Misuses Due to Misleading Charts

Example 116A   Details Magnified Out of Context

Fig. 117A shows the course of consumer prices from 1933 to 1953. The rise at the end of the chart can be made quite startling on a chart confined only to it, as in Fig. 117B. The omission of the zero line leaves room to magnify the vertical scale, in this illustration by a factor of 10. The horizontal scale has also been magnified, but only by a factor of 3, so the rise is 3½ times as steep in Fig. 117B as in Fig. 117A.

Example 116B   Perspective

Perspective diagrams are hard to interpret. Fig. 117C is supposed to depict the change in the national debt from about 1860 to the present time.[23] This presentation grossly distorts the amplitude of the recent fluctuations. The visual impression is that the debt in 1948 is about 10½ times the debt of 1920, but

---

[23] This is the cover design used by the Committee on Public Debt Policy for its *National Debt Series*, issued between World War II and the Korean War.

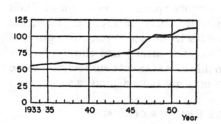

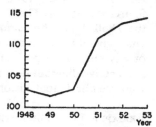

FIG. 117A. Consumer price index, 1933–53. (1947–49 = 100.)

FIG. 117B. Detail of Fig. 117A.

Source: U. S. Bureau of Labor Statistics, *Monthly Labor Review.*

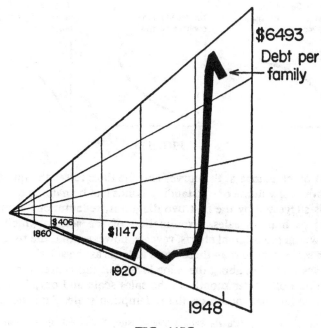

FIG. 117C

the ratio between 1948 and 1920 computed from the debt figures is only 5⅔. The 1948 figure appears to be about 63 times the 1860 figure, but actually was only 16 times it. Thus, the chart gives two to four times the legitimate impact. The purpose of any chart is to present the facts clearly and simply. Such a perspective diagram does neither. It is easy to suspect that those who use charts that distort may not have a good case.

EXAMPLE 118   DECEPTIVE CHANGES OF SCALE

Fig. 118 sketches the general appearance of a misleading series of charts relating to sales of U. S. Government Series E bonds in the period 1941–1944. It was presented as a model of what "a lively imagination in selecting and compressing data" can do.[24]

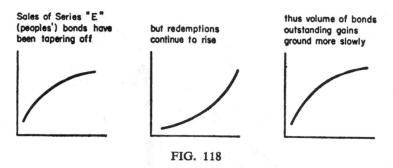

Sales of Series "E"
(peoples') bonds have
been tapering off

but redemptions
continue to rise

thus volume of bonds
outstanding gains
ground more slowly

FIG. 118

A quick glance at the curves and the titles raises the question of how the volume of outstanding bonds can be gaining at all if, as is suggested by the first two diagrams, redemptions are running as high as sales. Examination of the scales, which are shown on the original charts, reveals, however, that the redemption scale is more than three times as large as the sales scale. The curves all end at about the same level, but this represents 1,000 million dollars per month on the sales scale and only 300 million dollars per month on the redemption scale. And the third

[24] J. A. Livingston, "Charts Should Tell A Story," *Journal of the American Statistical Association,* Vol. 40 (1945), pp. 342–350.

scale is totally different, a logarithmic scale. If the data on volume outstanding are plotted on an arithmetic scale the resulting curve looks like Fig. 119, the final level being about 25 billion dollars. The absolute amount added to the volume of outstanding bonds is greater each month than the month be-

FIG. 119

fore. As a *percentage* of the total outstanding in the previous month, however, the increase in volume of outstanding bonds is decreasing. The logarithmic scale shows the latter—the percentage rate of growth. On a logarithmic scale, equal rises on the vertical axis represent equal *percentage* increases.

EXAMPLE 119   CARELESS SCALES AND LABELS

The following graph was intended to show that the steel industry was in much worse financial shape than was generally realized.[25] The point may or may not be well taken, but the chart exaggerates the effect of taxes upon profits.

In the first place, the base line is not zero, so a false impression is made of the amount of the profits, especially for 1945 when profits fall to the base line.

The title of the chart is "Profits after taxes," but the vertical axis measures profits after taxes not directly, but as a percent of total income, which differs in important respects. Dollar profits, which the casual reader may think he is reading from this chart, have actually gone up considerably.

[25] *Steelways* (published by the American Iron and Steel Institute), March, 1948, pp. 6–7.

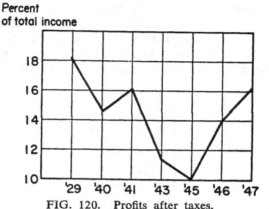

FIG. 120. Profits after taxes.

A final criticism is the manner in which the years are marked off on the horizontal axis. Certain years are simply omitted. The same distance is used to represent a one-year interval in three cases, a two-year interval in two other cases, and an eleven-year interval in one case. If the years 1929 to 1940 had been included year by year, an entirely different impression might have been created because of the tremendous rise in profits since the early thirties.

Conclusion

Our examples illustrate errors in the use of the basic definitions underlying an investigation, in the application of those definitions in the measurement or classification of individual people or objects, and in the selection of individuals for measurement. They also illustrate errors in the use of the resulting data, by making comparisons improperly, by failing to allow for such indirect causes of differences as heterogeneity of groups with regard to important variables, by disregarding the variability that is usually present even under apparently constant conditions, by technical errors, and by misleading verbal or graphical presentations.

The misinterpretations involved in most of the examples are simple and obvious. Some readers find them amusing; many are. Some find them distressing; many are that, too. Some find them irritating. Others find us irritating, for what they consider a negativistic, quibbling, and pettifogging attitude. Some become negativistic, quibbling pettifoggers themselves, conjuring up imaginary fallacies or exaggerating the consequences of real ones. Some readers are dismayed at the lack of systematic criteria which will automatically and authoritatively classify any particular use of statistics as sound or unsound.

Some readers, and we hope you are among them, recognize that each illustration is an example of a general type of fallacy. They realize that, easy as it may be to recognize and cope with such fallacies when they are exhibited caged here, it is not always easy to recognize them or cope with them in their native habitats. These people ponder each example carefully, however amusing, distressing, or irritating they may find it. They appreciate, too, that all these misuses, far from showing statistics to be useless, exemplify its usefulness; for each misuse represents an opportunity to find a sound basis for practical action or to obtain valid general knowledge.

For the statistician, not only death and taxes but also statistical fallacies are unavoidable. With skill, common sense, patience, and above all (as we said at the beginning of this chapter) objectivity, their frequency can be reduced and their effects minimized. But eternal vigilance is the price of freedom from serious statistical blunders.

Chapter 5

Samples and Populations

The following four examples illustrate a problem of interpretation which was not explicitly brought out in the preceding chapters.

EXAMPLE 123A   FAMILY INCOME, 1952

The Bureau of the Census reported that in 1952 one-third of the families in the nation had money incomes of less than $3,000, and one-third had money incomes above $4,500. A critic of the report wrote:

> Only 25,000 families were interviewed. . . . The main fault of the report is the limited scope of its available figures. 25,000 families are not a sound representation for the 46,000,000 families in the U.S. . . .

EXAMPLE 123B   RETAIL DRUG STORES

A student wrote as follows about a statistical study of retail druggists:

> It is based upon voluntary financial returns of 1,378 individual drug stores. The smallness of this number in comparison to the 55,796 drug stores shown in the 1948 Census of Business seems to invalidate this survey. To make broad generalizations about 56 thousand stores on the basis of data for only about 14 hundred stores is quite unjustifiable.

EXAMPLE 123C   CIGARETTE SALES

A restaurant attempted to evaluate the effect on its cigarette business of changing from sales by the cashier to sales by a coin-operated vending machine. The number of packages sold during the first month of machine sales was 51 percent less than during the last month of cashier sales. As a basis of comparison, sales for the same two months in two comparable restaurants were used. These showed decreases of 15 percent and 3 percent, respectively. As a result of this comparison, it was concluded that the "installation of the cigarette vending machine was detrimental to sales."[1]

EXAMPLE 124   RAILROAD TELEGRAPHER

During the presidential campaign of 1952, a prominent speaker said that since the period before World War II the cost of living for a particular railroad telegrapher had increased more than his income. It was implied that the same was true for the country as a whole.

In the first two examples, samples of 25,000 and 1,378 were considered too small for "broad generalizations." In the last two examples, samples of 3 and 1 were considered large enough. Which are right? This problem cannot be analyzed as readily by common sense as could the examples of Chaps. 2 and 4. It is true that a sample of 25,000 might be inadequate for some purposes. It is also true that very small samples might be adequate for some purposes—a single case showed quite conclusively that atomic bombs will sometimes explode. But common sense is a poor guide in determining how much evidence should be assembled to answer a given question, or whether the evidence actually presented is enough to justify conclusions someone has drawn.

The problem thus posed is a purely statistical one, the problem of *sampling*. In this chapter and the next one we shall introduce a few ideas that are basic to what comes later. At the risk of oversimplifying the idea of statistical sampling, we shall deliberately ignore almost all complications and ramifications that arise in practice, and concentrate on the heart of the matter.

What We Do Know and What We Want to Know

The two most fundamental concepts of statistics are those of a *sample* and a *population*.

A *sample* is often referred to as "the data" or "the observations": numbers that have been observed. The *population*, on

[1] Richard R. Still, "The Effect of an Automatic Vending Machine Installation on Cigarette Sales," *Journal of Marketing*, Vol. 17 (1952), pp. 61–63.

the other hand, is the totality of all possible observations of the same kind. In Example 123A, the sample consisted of income figures for 25,000 families. The population consisted of all the numbers that would be obtained by selecting indefinitely many families and measuring their incomes by the same methods that were used to select these 25,000 families and measure their incomes. Since in 1952 there were 46 million families in the country, it might seem that the population contains 46 million numbers. Actually, it contains many more, for the process by which a number is obtained for a given family might have yielded different numbers for that same family, depending on the interviewer used, the member of the family interviewed, the time of the interview, and so forth. The population in this example can be regarded as infinite.

A single population can give rise to many different samples; thus, the population is stable but samples vary. A central problem in statistics is to determine what generalizations about the population can be drawn from the one particular sample which is actually available in a practical problem.

We shall try to bring out some of these basic ideas by describing a sampling demonstration in which many samples were drawn from a single population.

Sampling Demonstrations

Apparatus and Method

A closed box containing an unknown number of red and green beads was used. The bottom of the box is a sliding panel with 20 depressions, into each of which one bead falls. This panel can be slid out of the box by pushing it with another panel which takes its place and keeps the remaining beads from escaping from the box. In this way, a sample of 20 beads is obtained each time a panel is removed.

The procedure in the demonstration was to mix the beads by shaking the box well, take a sample of 20 by sliding out the panel, record the number of red beads, and put the sample beads back into the box. This procedure was repeated 49 times, producing 50 samples in all.

The purpose of the demonstration was to illustrate the kind of results to be expected if repeated samples were drawn with the same initial conditions. Since the beads from one sample were returned to the box before another sample was drawn, the population remained the same. It was as if one person studied the population of beads by drawing a single sample of 20 and counting the number of red beads, a second person studied the same population, and altogether 50 people studied it independently by exactly the same methods of selection and measurement ("measurement" in this example being simply a matter of classifying each bead as red or not red and then counting the number of red beads in the panel). Had we not replaced the beads after each sample of 20, the contents of the box would have changed every time a sample was drawn. We wanted each sample to arise in the same way it would have in a practical situation in which we were taking only one sample.

The problem of the beads in the box is statistically equivalent to innumerable practical problems. For example, suppose there is a group of 700 income tax returns of which an unknown proportion are incorrect. From this population we might take a sample of 75 returns, audit them, count the number that are incorrect by more than a specified amount, compute the proportion incorrect in the sample, and apply this figure to the whole 700. Or suppose there is a population of 1,000 hospital patients with the same illness, of which an unknown proportion will respond favorably to a certain treatment. Or, again, suppose a meteorologist interested in rain-making wants to estimate the proportion of clouds of a given type which, under specified conditions, would produce rain within 20 minutes after seeding by a certain method. All of these situations, and many more, reduce to this: From a population in which an unknown proportion of the items have a characteristic in which we are interested, a sample is drawn and the proportion of the sample having that characteristic is determined. On the basis of what has been learned from the sample, a decision is made concerning the entire population. The consequences of this decision depend on what proportion of the population has the characteristic whose frequency in the sample was computed. The consequence of treating or not treating patients in a certain way, for example, depends on the proportion in the population of patients who will respond favorably to the treatment.

Let us introduce some symbols to represent the various quantities described in the preceding paragraph. This will make our further discussion briefer, more precise, and more general. It will also serve as an introduction—a gentle one, we promise—to the use of simple mathematical notation for statistical ideas.

The number of individuals in the population may be represented by N. For the case of the box of beads, N is unknown, but much larger than 20; for the income tax example $N = 700$; for the medical example $N = 1,000$; and for the cloud-seeding example $N = \infty$. This last, read "N equals infinity," means that no definite number, however large, will be as large as the number of possible reactions to seeding of all the clouds of the given type that ever have existed or ever will exist. The proportion of the items or individuals in the population that have the characteristic in which we are interested—redness for the beads, errors for the tax returns, and so on—is an unknown number between 0 and 1, inclusive; we may denote it by P. If the numerical value of P were known, it would determine the proper decision in the matter at hand and there would be no statistical problem and no need for a sample. Since P is not known, we draw a sample and base our decision on that. Let n represent the size of the sample, that is, the number of items from the population which are actually observed. For the beads, $n = 20$; for the tax returns, $n = 75$; and for the other examples the value of n was not stated, but would be known. Let X represent the number of items in the sample which have the characteristic in which we are interested. Finally, let p represent the proportion of the individuals in the sample having the characteristic in which we are interested; that is, $p = X/n$. For example, if 19 red beads are found in a sample of 20 from a population of 3,000, then $N = 3,000$, P is unknown, $n = 20$, $X = 19$, and $p = 19/20 = 0.95$.

Always distinguish clearly between P, the proportion in the population, and p, the proportion in the sample. The consequences of the decision depend on P, and if it were known we would know what decision to make. When P is unknown, we face the statistical problem of basing our decision on the incomplete information about P given by p.

For demonstration purposes, the box has several advantages over most practical situations. It is compact, and many samples can be taken in a short time. Furthermore, there is (except for the color blind) no trouble in distinguishing between a red bead and a green bead, but it may be difficult or time-consuming to distinguish a correct tax return from an incorrect one, a patient who responds favorably from one who does not, or even a cloud that rains from one that does not.

The demonstration which we will describe shows that samples vary, but it also shows that random samples vary according to a regular pattern. It would take many more than 50 samples to disclose the pattern perfectly, but 50 will suffice to show the general outline.

Preliminary Sample

Typically, in practical work we have only one sample. So we first examined one sample of 20 beads and considered it by itself, as if it were to be the only sample. In this preliminary sample, the number of red beads, X, was 13. Thus, the proportion p in the sample was $13/20 = 0.65$; that is, the sample contained 65 percent red beads.

With only this information available about the contents of the box, some might conclude that there are exactly 65 percent red beads in the whole box. Others might say that a sample of 20 is too small to tell anything about the proportion in the box. Both of these extreme positions are wrong. Clearly there is no basis for thinking that the sample is a perfect miniature replica of the population. If the true proportion, P, were 0.628, for example, 0.65 would be as near as the sample proportion, p, could possibly come in a sample of 20; but chance could lead to such results as 0.55, 0.70, and so on, even if P were 0.628. But, just as clearly, *something* has been learned from the sample. The possibilities that $P = 0$ or $P = 1$ have been ruled out completely. The possibility that $P = 0.1$ is not ruled out completely, but probably no one will take that possibility seriously, for if P were really as small as 0.1, the probability of getting as many as 13 red beads in a sample of 20 would be extremely small—it would happen, on the average, less than once in a hundred million times. Clearly, by this type of reasoning, some idea about *P* can be obtained from *p*.

Fifty Samples from Population I

The preliminary sample of 20 contained 13 red beads. But as we have seen, this is partly accidental. We have to judge it against the variation in the samples that would occur if the same conditions were repeated many times. Since this is a demonstration, rather than a real problem, we have an opportunity to find out something about "what might have been" if we "had it all to do over again." We did it all over again—not once, but 50 times. The results are shown in Table 129.

If these results are grouped in a *frequency distribution*, as in Table 130, there begins to emerge a general pattern of variation

TABLE 129

RESULTS OF 50 SAMPLES FROM POPULATION I

(20 Beads per Sample; Total of Sample Sizes, 1,000)

Sample Number	Red Beads	Sample Number	Red Beads	Sample Number	Red Beads
1	12	18	12	35	12
2	12	19	8	36	11
3	9	20	12	37	10
4	12	21	9	38	10
5	9	22	11	39	9
6	9	23	9	40	10
7	11	24	10	41	11
8	8	25	14	42	9
9	11	26	11	43	11
10	8	27	12	44	12
11	11	28	14	45	12
12	10	29	10	46	12
13	7	30	14	47	13
14	8	31	10	48	10
15	13	32	18	49	10
16	13	33	8	50	13
17	13	34	15	Total	548

for samples of 20 from this particular population. The bulk of the samples had 9, 10, 11, or 12 red beads. The preliminary sample we talked of earlier, with 13 red beads, is now seen not to have been one of the most usual results, though it is not particularly unusual either. At any rate, we would have been foolish to draw any conclusions from that preliminary sample without taking into account "what might have been"—that is, without taking into account the pattern of variation among samples which is revealed (partially) by Table 130.

TABLE 130

FREQUENCY DISTRIBUTION OF SAMPLE
RESULTS, POPULATION I

Number of Red Beads	Number of Samples
Less than 7	0
7	1
8	5
9	7
10	9
11	8
12	10
13	5
14	3
15	1
16	0
17	0
18	1
More than 18	0
Source: Table 129. Total	50

Fifty Samples from Population II

What appears in a sample depends on chance, as this demonstration reveals. But the pattern of chance variation in repeated samples depends on the population. This was illustrated by changing the contents of the box, thus forming a new population. The pattern of variability changes when the contents are changed. This fact, that the pattern of variability depends on the population, provides the possibility of making generalizations about a population from a random sample. The problem is one of stripping away the effects of chance, as far as possible, to see the reflection of the population in the sample.

The new set of beads was called Population II. The results of the 50 samples from it are given in Table 131A and the frequency distribution is shown in Table 131B.

TABLE 131A

FIFTY SAMPLES FROM POPULATION II

(20 Beads per Sample; Total of Sample Sizes, 1,000)

Sample Number	Red Beads	Sample Number	Red Beads	Sample Number	Red Beads
1	3	18	4	35	2
2	4	19	4	36	5
3	1	20	4	37	4
4	2	21	5	38	3
5	1	22	3	39	4
6	2	23	5	40	4
7	3	24	2	41	3
8	1	25	2	42	2
9	5	26	2	43	3
10	1	27	2	44	4
11	3	28	2	45	2
12	3	29	2	46	3
13	4	30	4	47	7
14	3	31	2	48	3
15	3	32	2	49	2
16	3	33	4	50	4
17	3	34	3	Total	152

TABLE 131B

FREQUENCY DISTRIBUTION OF SAMPLE RESULTS, POPULATION II

Red Beads in Sample	Number of Samples
0	0
1	4
2	14
3	15
4	12
5	4
6	0
7	1
More than 7	0
Total	50

Source: Table 131A.

The pattern of variability for the samples differs between Populations I and II in several respects, of which two are conspicuous: the patterns cluster around different values, and the tightness of the clustering differs. The samples from Population I cluster mainly from 9 to 12 beads, with approximately equal numbers above and below 11; those from Population II cluster mainly from 2 to 4, with approximately equal numbers above and below 3. A span of 12 (7 through 18) is necessary to encompass all samples from Population I, while a span of 7 (1 through 7) is sufficient for Population II. These two differences, in the location and in the dispersion of the patterns, can be seen in Fig. 133 which displays the results of all the samples from both populations.

A chart like Fig. 133 is an excellent means of recording data as they occur. It shows vividly the variation from sample to sample, yet it also indicates the general level about which the variation occurs. A frequency distribution like Tables 130 and 131B can be prepared easily from a chart like Fig. 133; for example, looking across at a certain horizontal level, say the one representing samples with 10 red beads, we see quickly that there were 9 such samples from Population I. Furthermore, if there were relations among the successive samples, such as trends or cycles, these would be apt to catch the eye. Our method of sampling in this experiment included precautions to preclude such relations, except such as are due to chance—for example, the run of 6 consecutive twos in the Population II (samples 24–29). We shall have more to say about the practical uses of such charts.

Conclusions from the Demonstrations

From the results of the sampling demonstration, two important conclusions stand out:

(1) Sample results vary by chance.

(2) The pattern of chance variation depends on the population.

These two facts correspond with the basic problem and the basic tool of statistical analysis. The basic problem is that a

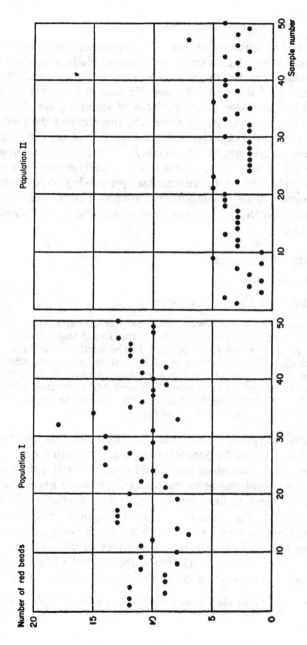

FIG. 133. Number of red beads per sample in 100 samples of 20 beads each. Source: Tables 129 and 131A.

sample is not a miniature replica of a population, so when decisions about a population are based on a sample it is necessary to make allowance for the role of chance in determining the characteristics of the particular sample that is available. The basic tool is knowledge of the patterns of sampling variability that result from various populations, and therefore of the probability of getting the observed sample from any of the different populations which might conceivably have been its source. Thus, given a sample, it is possible to say of certain populations that the sample might with "reasonable" probability have arisen in the normal course of sampling variability. It is "reasonable" to conclude that the population from which the sample came is one of these.

Variability of Samples

EXAMPLE 134   MOTIVES

Consider the following quotation:

> Interviews were performed with 20 persons, and motives extracted from the experience data. It was found that between 75 and 80 percent of these persons had a particular motive. Then another 20 interviews were added and still approximately 75 percent of the people had this particular motive. Additional interviews were added in blocks of 20 until 140 interviews had been taken. This particular motive still applied to approximately 75 percent of the 140 persons.[2]

There are several objections to this quotation—for example, the percent of a group of 20 people having a characteristic can not be between 75 and 80, since this would imply a fractional number of people—but the point pertinent here is the impression conveyed that *each* of the seven samples of 20 contained about 75 percent with the particular motive. Suppose that three-fourths of the beads in a sampling box like the one described earlier are red, and that successive samples of 20 are drawn. There will be much more variation in the percent of red beads from sample to sample than the quotation implies. In two series

[2] William A. Yoell, "How Big a Sample in Qualitative Research?" *Advertising Age and Advertising and Selling*, September, 1950.

of seven successive samples of 20, for example, we obtained the following percentages of red beads:

> First series: 75, 85, 75, 60, 70, 70, 65
> Second series: 55, 65, 85, 70, 60, 90, 80

Only two of our 14 samples produced exactly 75 percent, and only six from 70 to 80 percent. Actually, our 14 samples happen to show a little more variability than would be expected "on the average," but even on the average only about one sample in five would yield exactly 75 percent. Nearly half would be 65 or less, or else 85 or more.[3]

It is a fundamental fact that different results are obtained under apparently fixed conditions. Thus, holes drilled with a given drill will all have different measurements, even with the same operator and the same material. The holes may be similar in the sense that any differences among their dimensions are of no practical importance. If fine enough measurements are taken, however, they will always show some variation. Part of the difference in dimensions will be "real," and part may be due to inaccuracies in measurement. If the same hole is measured repeatedly by the same person or different persons, there will be some variation in the recorded measurements if the gradations are fine enough. Similarly, a litter of rats will exhibit varying individual growth despite the most carefully controlled heredity, environment, and measurement. We can regard this as a controlled "process," much as we regard the machine process that drills repeated holes. By a "controlled" or "apparently fixed"

---

[3] Another aspect of the quotation deserves comment. Cumulating the samples, as the author apparently did, introduces an artificial appearance of stability. Thus, our first series shows 75 percent for the first sample of 20, 80 percent for the first two samples, 78 percent for the first three, and so on. Cumulated in this way, our series (rounded to two figures) become:

> First series   75, 80, 78, 74, 73, 72, 71
> Second series: 55, 60, 68, 69, 67, 71, 72

There is an appearance of stability here, at least in comparison with the original series, but its results from the arithmetic of cumulating, not the stability of successive samples. For example, even if an eighth sample were to show 100 percent, either series would be raised only by 4. As the number of samples already taken becomes greater, the effect of the next sample on the over-all average becomes less. (To be specific, sample number k will change the previous averages by one kth as much as it differs from the previous average; thus, in our first series, sample number 3 with 75 percent is 5 below the previous average of 80, so it lowers the average by ⅝ to 78⅓, which is rounded off to 78.)

process we mean one in which the individual differences cannot be associated with identifiable or "assignable" causes, but rather are such as we observed in sampling from Populations I and II, which we ascribe to "chance."

Thus, the items in a population almost always vary among themselves. Even if there were no "real" differences, the process of measurement might introduce variability. It is this variability among the items of the population that leads to variation among samples, for the different samples include different sets of the population items. The greater the variation among the population items (that is, the less homogeneous the population), the greater, other things equal, will be the variability among samples. In the sampling box, for example, when one of the two colors predominates there is, in a sense, less variability in the population, hence less variability from sample to sample, than if the two colors are equally represented. This is illustrated by the fact that the probability that both beads of a pair will be the same color is higher the more one color predominates in the population. At one extreme, when there is no variability in the population—that is, the beads are completely homogeneous in color—there is no variation among the samples. At the other extreme, when the population is evenly divided, its variability is at a maximum and the variation from sample to sample is also at a maximum. That is why the samples from Population I were more variable than those from Population II.

In analyzing samples we will repeatedly fall back on the question, "Can the observed differences reasonably be explained by chance?" Only when the answer is "no" does the evidence imply assignable causes.

Reasons for Using Samples

It would have been possible to determine the proportion of red beads in the sampling box by counting the whole population, but in many situations this would be either impossible or impractical. It would be most impractical, for example, for a mail-order company to open every outgoing package to classify it as satisfactory or unsatisfactory, and thereby determine the proportion of its orders being filled correctly. In general, when observing an item destroys it, as in this case and in measurements of durability or breaking point and in tests of functioning on such items as fuzes or matches, inspection of the whole population is out of the question.

Even where complete inspection is possible, sampling may have economic advantages. Resources—materials, time, personnel, and equipment—constitute a limitation in any investigation, and it is necessary to balance the information obtained against the expenditure. It may be that measuring only a sample instead of the entire population results in a margin of potential error, known as *sampling error* (referring to error that in all probability *might* occur because of sampling, not to error that necessarily *has* occurred or will occur because of sampling), small enough for practical purposes—that is, small enough so that a reduction in this risk of error would not be worth the cost of achieving it by further observations. For example, if measuring the useful life remaining in telephone equipment on only three percent of the units gives a sampling error for the whole plant of less than 0.5 percent (as was true in one actual case), it would not ordinarily be worthwhile to measure more items unless the measurements were virtually costless—which, of course, they are not.

The concept of sampling error will require further clarification later; its nature is implied, however, by the following (true) statement about sampling error: In situations like that of the sampling box, involving determination of the proportion of a population having a specified characteristic, the sample proportion will be within 0.05 of the population proportion for at least 95 percent, on the average, of samples of size 404 or more.[4]

Two reasons have been given so far for using samples instead of complete surveys: that sometimes the measuring process destroys the items, and that the gain in accuracy from a complete survey may not be worth the cost. A third reason is that the

[4] If $P = 0.5$, the sampling variability of p will be greater than if P has any other value. (This was explained in the preceding section, on "Variability of Samples.") If P should be 0.5, samples of 404 would produce values of p between 0.45 and 0.55 (that is, within 0.05 of P) just 95 percent of the time, on the average. The farther P is from 0.5, the greater the percentage of the time that p will fall within 0.05 of P. To make our statement true for any value of P, we have allowed for the least favorable case.

individual measurements may not be as accurate for a complete survey as for a sample. A large number of measurements made hurriedly or superficially may not represent as much true information as a small number made carefully. In extreme cases, poor data can be so misleading as to be worse than no information at all. A rather paradoxical example of the effective use of samples is the Bureau of the Census' use of them to check on the accuracy of the census. Although sampling error is almost absent from the census, the nonsampling errors may be considerable—that is, such errors as those arising from failure to make questions clearly understood, from misrecording replies, from faulty tabulation, from omitting people who should have been interviewed. In the sample census, however, these nonsampling errors may be reduced enough to offset the sampling error, for it is cheaper and easier to select, train, and supervise a few hundred well-qualified interviewers to conduct a few thousand careful interviews than it is to select, train, and supervise 150,000 interviewers to conduct a complete census of the population.[5] Similarly, in measuring the useful life of the equipment in a telephone plant, the practical choice is not between measurements for a sample of the equipment and equally accurate measurements for all the equipment, but between fairly precise measurements of a sample made carefully by competent engineers, and crude measurements of the whole plant made hastily by less skilled people. Even in laboratory experiments in the sciences, the difficulties of precise measurement are often so great that it is better to reduce the number of items measured in order to take more care with the individual measurements.

A fourth reason for sampling is that a complete survey may be impossible because the population contains infinitely many items. The reactions of clouds to seeding is a case in point. Simi-

[5] The reader may wonder why, in view of the advantages of sampling, the entire population of the United States is enumerated completely every ten years. Aside from the over-riding fact that the Constitution requires this, perhaps the most important reason is that information is required for very small groups of the population—such as small towns, individual neighborhoods in cities, etc.—as well as for the country as a whole. Even so, however, about half the questions on the 1950 census were asked only of a sample—for some questions a 20 percent sample, and for some questions a 3⅓ percent sample (namely, a 16⅔ percent subsample of the 20 percent sample).

larly, in studying the effects of a medical treatment, the population ordinarily is all responses to the treatment that will ever occur with patients in a certain condition, an essentially infinite population.

A fifth reason for using samples is that for many data the population is inaccessible, and no more data can be had from it. This is particularly true of time series—historical records giving measurements of some phenomenon at various dates in history. Careful records of the level of Lake Michigan (and of each of the Great Lakes), for example, are available for each month from January, 1860 to the present. For studying the seasonal pattern of the lake's level—that is, the relation between the levels for the various months of the year—or for studying such other characteristics as cycles, trends, and extremes in its level, the records constitute a sample (as of the end of 1962) of 103 years. It may be possible to extend the record prior to 1860 by using less accurate or less systematic observations, but not much reduction in the sampling error can be expected this way except at the expense of introducing errors from the unreliability of the data; and, of course, nothing but watchful waiting can extend the record into the future. Nevertheless, the seasonal variations shown in the 103 years must be regarded as a sample, in that continuation of the same basic forces and processes does not produce an identical pattern of fluctuation each year. A person facing a problem for which the correct decision depends on the seasonal pattern needs to make allowances for the extent to which the 103 observations reflect sampling variation.

All five of these reasons for sampling fall under the general principle that there comes a point beyond which the increase in information from additional observations is not worth the increase in cost.

Conclusion

Population is an abstract concept fundamental to statistics. It refers to the totality of numbers that would result from indefinitely many repetitions of the same process of selecting objects, measuring or classifying them, and recording the results. A population is, thus, a fixed body of numbers, and it is this general body of numbers about which we would like to know. What we actually know is the numbers of a *sample*, a group selected from the population. Because the numbers in the population almost always vary, both inherently and through variations in measuring and recording, the results of a sample depend on which numbers from the population are included in the sample. In generalizing from the sample to the population, therefore, allowance must be made for the fact that the sample results are partly fortuitous. This allowance is made by considering the pattern of sampling variability, called the *sampling distribution*, of sample results, that is, by considering the various samples that could occur from any particular population and their respective probabilities of occurring. Some populations are then seen not to be likely to produce a sample such as that observed, and others to be likely to produce it. The population from which the sample came is inferred to be one of the latter.

The principal reasons for using samples instead of observing the whole population are associated with the fact that once the sample attains a certain size, additional observations will not reduce the *sampling error*, that is, the allowance for sampling variability in the conclusions, enough to be worth the additional cost.

Chapter 6

Randomness

Meaning of Randomness

THE POINT stressed in Chap. 5 was *sampling variability*—its pattern and the relation of this pattern to the population from which the samples originate. An equally important matter illustrated by the demonstrations is *randomness*. The method of sampling we used is an example of random sampling.

"Random" as used in statistics is a technical word; it has a meaning different from the one given it in popular usage. When a sample is called "random," this describes not the data in the sample, but the process by which the sample was obtained. Thus, randomness is a property not of an individual sample but of the process of sampling, just as in a game of cards a fair hand is not one in which the cards have certain values, but one dealt by a certain process. In fact, what in card games is called a fair hand is precisely what in statistical terminology would be called a random sample of the deck.

A sample of size n is said to be a *random sample* if it was obtained by a process which gave each possible combination of n items in the population the same chance of being the sample actually drawn. Thus, in our demonstration, the thorough shaking of the box before each sample was drawn was intended to give each possible set of 20 beads the same chance as any other set of falling into the 20 holes in the panel. A primitive way to achieve randomness is to assign the numbers 1 to N to the N items in a population, write the numbers on cards or chips, thoroughly mix them, select a set of n, and then use as the sample the items corresponding to these n numbers. This way of achieving randomness is mentioned here primarily to clarify the meaning of randomness. Actually, these mechanical manipulations are surprisingly difficult to perform reliably, so in practice use is made of tables of random numbers that have been prepared for this purpose.[1]

Reasons for Randomness

Nonstatisticians usually assume that the importance of randomness arises from the "fairness" and lack of bias with which such samples represent the population. This is important, of course, but of more importance is the fact that *the pattern of sampling variability for any population is known if, but only if, the sampling is random.* As we said in drawing conclusions from the sampling demonstrations described in the preceding chapter, the basic tool of statistical analysis is knowledge of the patterns of sampling variability that result from various populations. This knowledge can be obtained only through the laws of mathematical probability, and these laws apply only to random samples. Thus, only random samples permit objective generalizations from the sample to the whole population. Two competent statisticians will reach similar conclusions from a given sample if it is known to be random and if they have agreed on methods of analysis; furthermore, they will agree on objective statements about the confidence to be put in their conclusions. They need not argue intuitively and interminably whether 25,000 items are enough or 3 too few to support a given generalization. The statistician thus depends on the fact that the pattern of variability of random samples from any population can be determined through the mathematical laws of probability. He not only recognizes sampling variability, he exploits it.

Perhaps the reason nonstatisticians tend to think of randomness only in terms of fairness is that this is its most obvious role in card games. Even in these games, however, knowledge of "the probabilities," as it is sometimes expressed, or of "the sampling distribution," as a statistician might express it, is

[1] The largest and best of these tables is that of The Rand Corporation, *A Million Random Digits with 100,000 Normal Deviates* (Glencoe, Illinois: Free Press, 1955). A smaller but also excellent table is the *Table of 105,000 Random Decimal Digits* issued free by the Interstate Commerce Commission, Washington (Statement 4914, File No. 261-A-1, Bureau of Transport, Economics and Statistics).

typically essential to effective play, even if one is assured his fair share of desirable cards.[2]

To repeat: Randomness is important in statistics primarily because if a sample is random, but not otherwise, the mathematical laws of probability are applicable and make it possible to know the patterns of sampling variability in terms of which the sample must be interpreted.

Randomness vs. Expert Selection

Almost any sampling method will have some pattern of variability, but random samples are the only ones that have a *known* pattern of variability. Consider an expert, whether he is an expert at judging the proportion of red beads in a box or at judging the proportion of Democratic votes in an election. This expert will not, of course, claim to specify the true proportion precisely. His figures are subject to sampling variability. But there is no way of knowing the pattern of variability in his method.[3]

This is not to disparage experts and expert judgments. It is better to rely on expert judgment for most everyday problems, than to make statistical studies of every question that arises. But when a statistical study is made, it should be an independent, objective *statistical* study, which may or may not confirm the expert's judgment. If the method is a mixture of statistics and expertise without the use of randomization, the result will be just one more huff or puff on a windmill which is probably spinning too freely already.

[2] This point is elucidated in a book by S. W. Erdnase describing methods of cheating at cards. He points out that simply changing the probabilities gives a considerable advantage to the person who knows the true probabilities himself and can profit by others' miscalculations. S. W. Erdnase, *Artifice, Ruse, and Subterfuge at the Card Table: A Treatise on the Science and Art of Manipulating Cards* (Chicago: F. J. Drake and Co., 1905). The cover bears the title *The Expert at the Card Table.*

[3] Before the sampling demonstrations described in Chap. 5 were actually done, several boxes of red and green beads were shown to one of the observers, and he was asked to estimate the proportion red by eye. Actually all the boxes had the same proportion red as Population I; his estimates were not all the same, but their average was quite close to the correct proportion for Population I. When this same "expert" (if we may take the liberty, for purposes of exposition, of qualifying him too easily as such) was shown a series of boxes all having the same proportion red as Population II, however, his average differed greatly from the correct proportion.

If the sample for a statistical study is selected according to expert judgment it *may* give better results than if it is selected according to statistical principles—provided the expert is so expert that a statistical study was not needed anyway. But such results do not reinforce, they only reiterate, the expert's judgment. It is important to see the contrast between statistical method and expert judgment. When we select data solely by judgment, expert or otherwise, we rely on a man; when we select data by random sampling, we rely on a method. The purpose of collecting facts is to give them full opportunity to support or contradict judgment, thereby adding to the knowledge available.

EXAMPLE 144 SAMPLING CASTINGS

An engineer has told us of a sampling blunder in his company which stemmed from using a convenient but nonrandom sampling method. The company had found from past experience that about 90 percent of certain castings it was buying were defective. The defects usually showed up only after some machining had been done. One supplier claimed that he had developed a new method which would virtually eliminate defective castings.

When the first new lot was received it was decided to take a sample of the lot and have the sample items X-rayed before any machining was done. A sample of 20 castings was taken from the top of the box containing the entire lot and the X-ray inspection did show a great improvement in quality (actually no flaws were detected). On the basis of this the lot was accepted.

The lot was machined and 75 percent of the castings had to be scrapped. Subsequent inquiry showed that by an error of the supplier, the box was filled mostly with castings from the old method, with the new ones only on top.

Had the 20 castings been chosen randomly from the entire box, not just the top layer, there would have been only one chance in a trillion of finding no defective castings if the lot contained 75 percent defectives.

There are, to be sure, circumstances in which nonrandom sampling may be appropriate. (1) Random selection of samples is often more costly than nonrandom selection. This cost argument is not always as valid as it may seem, though, partly because the cost may represent ineptness at random sampling, and partly because random samples may give more valuable results. To put it another way, results of given value may be obtainable with smaller samples if sampling is random; indeed, they may be unobtainable with nonrandom samples of any size. (2) There may be occasions when only very few items can be included in a sample, as when an intensive study of cities is to be made, and even two cities would be too expensive, or when there is only enough of a new drug to treat, say, five cases, and the delay involved in getting more would be prohibitive. In these instances, generalization from the sample to the population will be essentially a matter of judgment anyway, since even a random sample of 1 or 5 will probably not alone justify any but foregone conclusions; so it is best for the expert who will have to make the judgment to select the cases in whatever way he judges will best illuminate the issue in his own mind. (3) Again, the argument that particular nonrandom methods of sampling have led to valid results in a certain kind of problem in the past always deserves serious consideration—though sooner or later such methods usually produce fiascos, as in the case of the *Literary Digest* presidential poll made in 1936 by methods that had proved successful in the previous four elections, or the case of the Gallup, Roper, and other presidential polls made in 1948 by methods that had proved successful in the previous three elections.[4] (4) Another situation where nonrandom sampling is appropriate is where only certain data are accessible, as in studying trends in the frequency of mental disease or the standard of living in 19th-century England. (5) Finally, random sampling—and even sampling of any kind— may be inappropriate where the object is to locate specific individuals, for example, the particular blood specimens with posi-

[4] Other factors than the sampling methods contributed to these fiascos.

tive reactions to a test for communicable disease, or the specific income tax returns with errors.

In all these situations except the last, however, an inverted question should be asked: Of what population are the data a random sample? That is, if whatever process produced the data were repeated indefinitely often, what population would it generate? The relation between this sampled population and the target population in which we are interested is then a question for expert consideration. Goldhamer and Marshall, for example, considered such a question in the study of mental disease that we discussed in Chapter Three. They would have liked to study the onset of all cases in the United States, but had to study instead hospital first admissions in Massachusetts; but they judged trends shown by these data to be satisfactory representations of trends in the onset of all cases. As a matter of fact, this inverted question—What population was actually sampled?—should be raised even where it has been possible to make a conscientious attempt to obtain a random sample from a clearly specified target population, for the best laid sampling plans (like other plans) can seldom, if ever, be executed perfectly. Subjects refuse to co-operate, animals die, machines break down, experimental materials are interchanged, etc.

Though randomization and expert selection are incompatible, this is by no means true of randomization and expert judgment about sample design. There are sound statistical methods for getting the best out of expert judgment without sacrificing the advantages of randomization. Suppose, for example, that we are planning a study of employee attitudes toward a firm, and an expert tells us that these attitudes vary with sex, race, union membership, department, and length of service. We could use the expert's judgment by *stratifying* our population—dividing it into a number of smaller populations (called *strata*) that are relatively homogeneous with respect to sex, race, etc.—and then sampling at random from each stratum. Furthermore, expert judgment is, as we emphasized in discussing the study of trends in the frequency of mental disease (Chapter Three), essential in selecting problems, deciding what to measure, and interpreting the findings of any statistical study.

Probability Samples

What we have described as random sampling is sometimes called *simple random sampling*, to distinguish it from various more elaborate sampling procedures, such as the stratified sampling just mentioned. All of these more elaborate sampling procedures are based ultimately on samples that are random in the sense we have described. The essential thing, as we have seen, is that the laws of mathematical probability should govern the pattern of sampling variability—that is, the *sampling distribution*—of such samples. A broader term, *probability sampling*, is used to describe any sampling process in which randomness enters at some stage in such a way that the laws of mathematical probability apply and provide the sampling distribution needed for interpreting a sample.

A probability sample of size n is one for which each set of n items in the population has a known probability of being the set chosen for the sample. More generally, the actual probability of selection for the sample need not be known; knowledge of the relation of this probability to the proportion of such samples in the population is sufficient. In the special case of a simple random sample, the probability of being the set chosen for the sample is the same for each set of n items in the population.

This chapter, confined to basic ideas, is not the place to discuss more elaborate probability sampling methods. Let us, however, briefly recall Example 99B, which dealt with estimating the number of children per family. The fallacy of this example can be viewed as one of analyzing as a simple random sample what is in fact a more complicated type of probability sample.[5] Here the chance of a family being included in the sample is not the same for all family sizes, so we do not have a simple random sample. But since we know the relation be-

---

[5] In Chap. 4, we interpreted the fallacy as one of failing to get a simple random sample when that was intended. The point is that the method of sampling and the method of analysis do not jibe. This can be interpreted either as the wrong analysis for the sampling method or as the wrong sampling method for the analysis; but it should be distinguished from sampling in such a way that the data are totally unanalyzable.

tween the number of children in a family and its chance of being included in the sample, we do have a probability sample. In fact, the probability of a family's being included is simply proportional to the number of children, and this knowledge makes it possible to estimate the average number of children per family free of the bias discussed in Chapter Four.[6]

Law of Large Numbers

Consider for a moment the 1,000 beads that were drawn from Population I as if they were all one sample from a very large population. We see from Table 129 that this sample shows 548 red beads, or a sample proportion of 0.548. In interpreting this result we must recognize, however, that a second sample of 1,000 would almost surely yield a different result, a third sample still a different one, and so on. Sampling variability could be demonstrated for samples of 1,000 just as it was for samples of 20, and from that point of view the sample of 1,000 reported in Table 129 now plays the same role as the preliminary sample of 20 discussed in Chapter Five. And if we were to proceed to draw 50 samples of 1,000, all beads drawn could be regarded together as one sample of 50,000, the interpretation of which would have to take account of the pattern of variability in samples of 50,000.

But the laws of probability assure the following: The probability that p will be within a given range of P is greater for samples of 100 than for samples of 20 from the same population, and still greater for samples of 1,000. For example, in samples of 20 from a population in which $P = 0.50$, about half the time p would be farther than 0.05 from P. In samples of 100 from the same population, deviations of p from P exceeding 0.05 would occur only a little more than one-fourth of the time. In samples of 1,000, such deviations would occur only about once in 1,000 times, and so on. This illustrates one aspect of what is called "the Law of Large Numbers." The larger the

---

[6] In effect, we simply disregard half the twos reported, two-thirds of the threes, three-fourths of the fours, and so on, and average the remaining observations. More precisely, we take a weighted mean of the data, weighting each observation by its own reciprocal.

samples, the less will be the variability in the sample proportions. Tosses of pennies illustrate the same thing. If a fair coin is tossed 50 times, the proportion of heads may well be as little as 0.4 or as much as 0.6. But if a fair coin is tossed 5,000 times, the proportion of heads is unlikely to fall outside the range 0.48 to 0.52.

There is another aspect of this law which must be borne in mind, and which can also be illustrated by coin-tossing. If two men match pennies repeatedly, the Law of Large Numbers does not guarantee, as many people think, that they will break even. The truth is that one of them will go broke. This is obvious if both players start with fortunes of only a penny, fairly plausible if they start with only a dime. It may seem less plausible as the fortunes increase, but nevertheless, the principle is true no matter how large the fortunes—though the time required becomes astronomical as the fortunes become even moderate. As the number of tosses increases, the number of wins for the two contestants may confidently be expected to diverge by larger and larger *amounts*, though by smaller and smaller *proportions*. A divergence of 500 wins would be highly improbable in 5,000 tosses, but could easily happen in 500,000 tosses; an excess of the larger number of at least 50 percent of the smaller number is highly probable (in fact, certain) in 5 tosses, but highly improbable in 5,000.

The arithmetic by which the absolute discrepancy can increase while the percentage discrepancy is decreasing deserves an illustration. Suppose 25 tosses of a fair coin show 15, or 60 percent, heads. This is an excess of 2½, or 10 percent of the number of tosses, above the expected even division. Now suppose 100 tosses show 55, or 55 percent, heads. This is an excess of 5 above the expected number, or twice as many as in the sample of 25, but an excess of 5 percent, or one-half as large a percentage as in the sample of 25. In general, if the departure of the number of heads from expectation increases, but less than in proportion to the number of tosses, the departure of the percentage of heads will decrease.

The naive conception of these principles, commonly spoken of as the "law of averages," is sometimes taken to imply that tosses of a fair coin approach half heads and half tails because after an unusual number of heads it is more likely that tails will turn up. This is not true, since each toss is independent of those before and after. The following anecdote illustrates the same point:

EXAMPLE 150   SEQUENCE OF BOYS

When 18 boys were born consecutively in five days recently at Walther Memorial Hospital it was generally assumed that a run of girl babies would begin. Since the run of 18 boys ended with the birth of a girl last Friday this is how the new arrivals have been recorded.

In the next six births, five were boys. Then four boys in the next six births. Again, four boys in the next five births. From Midnight Tuesday—five more boys and only one girl. Since October 31: 36 boys, 6 girls.[7]

As Tippett has put it, the Law of Large Numbers works by its "swamping" effect rather than by compensation.[8] An unusual result that produces 50 too many heads in 1,000 tosses, for instance, will not be perceptible in the proportion of heads after 99,000 more tosses, unless a similar excess occurs repeatedly.[9] But the basis of the Law of Large Numbers is that for an improbable event to occur n times is improbable to the nth degree.

Thus, while samples vary, averages and proportions vary less in large samples than in small samples from the same population.

Statistical Control

One reason for constructing Fig. 133 in the way we did is that it resembles charts called *control charts* that are often useful in

[7] *Chicago Daily News,* November 10, 1949.
[8] L. H. C. Tippett, *Statistics* (New York: Oxford University Press, 1943), p. 87.
[9] If 1,000 tosses show 550 heads, the percentage is 55. Assume that the next 99,000 tosses show 49,500 heads, or exactly 50 percent. Then the 100,000 tosses have shown 50,050 heads, or 50.05 percent. The effect of a large excess in a single thousand is thus "swamped," not compensated for. The other 99,000 tosses, of course, will not show exactly 50 percent heads, but they are as likely to show less as to show more if the coin is really fair. At any rate, whatever discrepancy from 50 percent there is in the whole 100,000 will be due scarcely at all to the first 1,000.

practical situations. All repetitive processes—no matter how carefully arranged—are accompanied by variability that cannot be explained by "assignable" causes. The process is said to be "in control" when the pattern of this variability is like that of independent, random samples from the same population—as in our experiment with the sampling box. Suppose we know

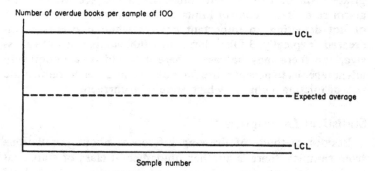

FIG. 151. Outlines of a control chart.

from past experience that a process, when in control, leads on the average to a certain proportion, P, of items having a certain characteristic. For example, in a certain library at a certain season the average proportion of books not returned by the date due may be known and stable. Then we can plot a control chart resembling Fig. 151, on which are to be plotted the results of samples for different days—say the number of books not returned among a sample of 100 due on a given day. UCL denotes "upper control limit"; LCL denotes "lower control limit."[10] These limits are usually set so that, if the process is in control, only about 3 plotted points in 1,000, on the average, will fall outside the limits. When a point does fall outside the limits, therefore, it is a signal to look for a change in the process for which, presumably, some explanation other than chance

[10] Statistical control limits should be carefully distinguished from "specifications" or "tolerance limits" which are commonly used in manufacturing. The latter usually refer to what the process should do; control limits are based entirely on what the process actually has done.

can be assigned. In the library example, there might, for instance, have been an increase in thefts or in errors in checking off returned books.

If the process is in control, there will be false alarms from 3 of every 1,000 samples, on the average. If this is too many, the statistical control limits can be set farther apart; the appropriate distance can be determined from the acceptable false alarm rate. If the control limits are set too far apart, the risk of not detecting an important change when it occurs is increased. Typically, 3 false alarms in 1,000 samples is not excessive, and there may be cases where it would be economically advantageous to accept more false alarms in order to reduce the risk of missing a change which should be corrected.

Statistical Description

Besides methods of drawing inferences about populations from samples, there is another fundamental class of statistical methods. It relates to the problem of description. The task of organizing and summarizing a particular body of data that has actually been collected is a problem in *statistical description*, in contrast with the task of drawing conclusions from data actually on hand about a larger body of data that have not been completely collected, which is a problem of *statistical inference*. These two problems are by no means completely separate, but it is feasible and useful to discuss them separately.

The value of skillful statistical description is suggested by Winston Churchill in a memorandum written while he was First Lord of the Admiralty in 1939–40:

> Surely the account you give of all these various disconnected Statistical Branches constitutes the case for a central body which should grip together all Admiralty statistics, and present them to me in a form increasingly simplified and graphic.
>
> I want to know at the end of each week everything we have got, all the people we are employing, the progress of all vessels, works of construction, the progress of all munitions affecting us, the state of our merchant tonnage, together with losses, and numbers of every branch of the R. N. and R. M. The whole should be presented in a small book such as was kept for me by

Sir Walter Layton when he was my statistical officer at the Ministry of Munitions in 1917 and 1918. Every week I had this book, which showed the past and the weekly progress, and also drew attention to what was lagging. In an hour or two I was able to cover the whole ground, as I knew exactly what to look for and where. How do you propose this want of mine should be met?[11]

We have already given one illustration of the process of statistical description when we prepared Table 130 from Table 129. The data of Table 129 are difficult to assimilate. About all we see is that the number of red beads varies, and usually begins with the digit "1." Table 130, however, presents the data in a way which immediately brings out the pattern in the variation and considerably sharpens our appreciation of the range in which most of the results lie. This is accomplished at the expense of obscuring the sequence in which the results occurred, the one thing that Table 129 does show, but our purpose in Table 130 was to highlight the pattern and shade the irrelevant and distracting features of the data. Fig. 133 is another description of the same data, one readily grasped by the eye, which brings out fairly well the pattern shown in Table 130 while also showing the sequence of occurrence, which would be important if control were in doubt.

Much of the most useful work done by statisticians consists in simply arranging masses of data so that they are comprehensible, or so that they focus attention on patterns and relations that are important, or else in summarizing masses of data by a few significant measures, for instance, an appropriate average. In Example 31A the principal contribution made by the statistician was in transcribing figures from the standard record forms and presenting them on a chart showing the relation between flying hours since overhaul and number of failures. Similarly in Example 31B (Merchant Ship Losses in Relation to Convoy Size), once the data had been properly organized and presented the conclusion was fairly clear.

Successful statistical description, like most successful statisti-

[11] Winston Churchill, *The Gathering Storm* (New York: Houghton Mifflin Company, 1948), p. 730.

cal work, depends greatly on knowledge of the subject matter. Mere manipulation of figures or preparation of standard tables and graphs is seldom fruitful unless guided by a clear conception of the subject matter and of what relations would be worth looking for. To a considerable extent statistical description is an art, rather than a science. As with other arts, however, there are certain basic techniques whose mastery is necessary, though not sufficient, for success. The brevity of our discussion of statistical description here, in this chapter on basic ideas, should not be taken to measure the relative importance of descriptive statistics. Rather, it reflects the fact that the field of descriptive statistics is not dominated by a few broad and pervasive principles as is analytical statistics, but is essentially a collection of techniques.

The remaining chapters emphasize statistical description, with some related discussion of sampling and inference. That is, they consider mainly the problems of describing a particular body of data already on hand, rather than the more technical problems of how to plan the collection of, and draw inferences from, data.

Conclusion

In drawing conclusions about a population from a sample, allowance for the fact that the sample results are partly fortuitous is made by considering the pattern of variability in samples. This requires deducing from any assumed population the sampling distribution that would result if samples were drawn from it. This is possible if, but only if, the sampling is *probability sampling*. A probability sample of size *n* is a sample selected from the population in such a way that there is a known probability of selection for every set of *n* numbers in the population, or at least that there is a known relation between the probability that this set will constitute the sample and the proportion of all sets of *n* in the population that contain the same numbers. A *random sample* (or better, a *simple random sample*, since frequently "random sample" is used in the sense of "probability sample" here) is one in which each possible set of *n* values in the population has the same probability that it will constitute the sample.

The distribution of samples depends on the nature of the population. For a given sample size, samples will be more variable the more heterogeneous the items in the population. And the *Law of Large Numbers* tells us that for a given population, the variability of sample averages and proportions will be smaller, the larger the sample.

A direct application of these ideas is the *statistical control chart*. The results of successive samples are plotted on this chart, and as long as the same population is being sampled virtually all of the sample results will lie between two limits, called *statistical control limits*. When a result falls outside the limits, it is evidence that there has been some change in the underlying population, for the sample is not within the probable range for the original population.

In contrast with *statistical inference*, which deals with methods of drawing conclusions or making decisions about populations on the basis of samples, *statistical description* deals with methods of organizing, summarizing, and describing data. Since the purpose of description is to prepare the way for inference, the basic ideas of inference have been outlined in this chapter and the preceding chapter as essential background for later chapters.

Chapter 7

Observation and Measurement

WHAT RELATION is there between the real world and the mere numbers that the statistician deals with? This is one of the most obvious, yet one of the most frequently overlooked, questions to be raised about any statistical study.

Strictly speaking, statistical analysis deals not with, say, the population of an area but simply with a set of numbers acquired in a certain way which it happens to please someone to designate by demographic terms; it deals not with the value of a certain plant and its equipment but with a set of numbers to which someone has attached words relating to plant and equipment; it deals not with the frequency of red beads in samples of 20 but with a set of numbers, produced by a certain process, which have been described in terms of red and non-red beads. In short, statistical analysis deals with numbers produced by certain operations, and strictly speaking its conclusions relate to the processes producing the numbers. But of course the interest the numbers have arises from their association with—their measurement of—things in the real world. In a sense, the relation between the numbers and the real world is not a problem for the statistician but for the subject-matter specialist—the engineer, sociologist, physicist, epidemiologist, etc. The meaning of the statistician's work is so completely dependent upon the meaning of the numbers with which he works, however, that he cannot wisely draw a sharp line between method and substance.

These ideas are brought out vividly in the following quotation:

> When we speak of "observing" business cycles we use figurative language. For, like other concepts, business cycles can be seen only "in the mind's eye." What we literally observe is not a congeries of economic activities rising and falling in

unison, but changes in readings taken from many recording instruments of varying reliability. These readings have to be decomposed for our purposes; then one set of components must be put together in a new fashion. The whole procedure seems far removed from what actually happens in the world where men strive for their livings. Whether its results will be worth having is not assured in advance; that can be determined only by pragmatic tests after the results have been attained.

This predicament is common to all observational sciences that have passed the stage of infancy. An example familiar to everyone is meteorology. The layman observes the weather directly through his senses. He sees blue sky, clouds, snow, and lightning; he hears thunder; he feels wind, temperature and humidity; at times he tastes a fog and smells a breeze; he sees, hears, and feels storms. The meteorologist can make these direct observations as well as a layman; but instead of relying upon his sense impressions he uses a battery of recording instruments —thermographs, barographs, anemometers, wind vanes, psychrometers, hygrographs, precipitation gauges, sunshine recorders, and so on. That is, he transforms much that he can sense, and some things he cannot sense, into numerous set of symbols stripped of all the vivid qualities of personal experience. It is with these symbols from his own station and with similar symbols sent to him by other observers dotted over continents and oceans that he works. . . .

All of us can observe economic activities as easily and directly as we can observe the weather, for we have merely to watch ourselves and our associates work and spend. What we see in this way has a wealth of meanings no symbols can convey. We know more or less intimately the hopes and anxieties, efforts and fatigues, successes and failures of ourselves and a few associates. But we realize also that what happens to us and our narrow circle is determined largely by what is being done by millions of unidentified strangers. What these unknowns are doing is important to us, but we cannot observe it directly.

A man tending an open-hearth furnace has a close-up view of steel production. But what he sees, hears, smells and feels is only a tiny segment of a vast process. He works at one furnace; he cannot see the hundreds of other furnaces in operation over the country. And smelting is only one stage in a process that includes mining and transporting iron ore, limestone, coal, and alloys; the getting of orders for steel, the erection of plants, and the raising of capital; importing and exporting, hiring and training workers, making and selling goods that give rise to a demand for steel, setting prices, and keeping account of outgo and income. No man can watch personally all these activities. Yet those engaged in them and in the activities dependent on the steel industry need an over-all view of what is happening. To get it they, like meteorologists, resort to the use of symbols that bear no semblance to actual processes and that are compiled mainly by other men.

For the intermittent process of making steel in a furnace with its heat and noise, its dim shadows and blinding glares, they substitute a column of figures purporting to show how many tons of steel ingots have been turned out by all the furnaces in a given area during successive days or weeks. That colorless record gives no faintest idea of what the operation looks like or feels like; it does not tell whether the work is hard or easy, well or ill paid, profitable or done at a loss. It suggests continuous operation, which is achieved at no furnace. It hides differences of location and types of product. And it separates the one act of turning out tonnage from all the other activities with which it is interwoven. Many, though not all, of these interrelated changes are likewise recorded in columns of figures; but each record is as devoid of reality and as divorced from its matrix as the record of tons produced.[1]

The Relationship Between a Number and the Real World

In considering any statistical analysis, it is always a good idea to stop and think about the relationship that the numbers bear to the subject and questions in which we are interested. Numbers by themselves have no meaning or significance; their significance depends on the circumstances and events that gave rise to them.

Consider a number labeled "number of passenger automobiles produced in the United States during the week ended October 28, 1961." Did someone stand at the end of the assembly lines tallying finished units? If so, was every line included, and was the tallying perfectly accurate? And how were all the separate counts compiled for the total figure? Perhaps each manufacturer reported his own figures, some based on production schedules, some on the number of engines sent to the assembly line, and some on the number of automobiles not only off the assembly line but also approved on a final inspection

[1] Arthur F. Burns and Wesley C. Mitchell, *Measuring Business Cycles* (New York: National Bureau of Economic Research, 1946), pp. 14–16.

process. What of cars partially produced during the week; do two half-finished cars count as one, and if so, when is a car half-finished? What of production of parts for shipment abroad unassembled?

Far more complex are the operations resulting in the number labeled "index of consumer prices, October 1961," or that labeled "velocity of light," or that labeled "safe concentration of carbon dioxide," or that labeled "patellar reflex reaction time."

In each case, there is a sequence of operations resulting in a number, and the significance of the number depends on those operations. Thus, in the case of passenger car production, if there is a tally of each unit as it rolls off the line, the resulting figures are only partially a measure of productive activity, for they do not reflect plant maintenance, manufacture of parts, and other productive activities except insofar as these other productive activities are closely correlated with the number of units coming off the line. If a line has just been opened, there may be a large amount of productive activity on incomplete units that is not reflected in the tally at the end of the line; or when the line is about to be closed, there may be tallies recorded for cars on which only a few final steps were taken during the relevant period. On the other hand, if completed units leaving the assembly line is precisely the quantity of interest, tallies at the end of the line will provide excellent measurements. Similarly, the number of people admitted to mental hospitals may or may not be a good measure of the prevalence of mental illness, and the number of crimes reported by the police may or may not be a good measure of the prevalence of crime.

Here are some examples in which numbers did not mean what they seemed to:

EXAMPLE 160   MUSEUM ATTENDANCE

We hear of a museum in a certain Eastern city that was proud of its amazing attendance record. Recently a little stone building was erected nearby. Next year attendance at the museum mysteriously fell off by 100,000. What was the little stone building? A comfort station.[2]

[2] *This Week*, April 17, 1948.

EXAMPLE 161A   NITROGLYCERIN

Fuel for rockets was being produced in fairly large batches.
The percentage of nitroglycerin in the product was supposed
to be of crucial importance to safety and to performance.
Therefore, an elaborate chemical analysis was made on each
batch. This analysis produced a number for each batch, labeled
"percent nitroglycerin," and the batch was accepted or rejected
on the basis of the number. Large amounts of material were
rejected and reprocessed, at considerable cost. Then a statistical
investigation was made of the relationship between the reported
percentage of nitroglycerin and actual performance of the fuel.
No difference in performance was observed between accepted
and rejected batches. This meant that the chemical analysis
must have been wrong, or else the theory about the relation
between nitroglycerin content and performance. It was discovered that the chemical analyses, despite their impressive
appearance, were producing numbers that bore little relation
to the actual percentage of nitroglycerin. In fact, the product
was less variable than the chemical analysis. The numbers on
which decisions were being made bore too rough a relationship
to performance.

EXAMPLE 161B   LIFE-RAFTS

Rubber life-rafts were tested individually for impact resist-
ance by dropping them from the height of a ship's deck into a
pool of water. Only rafts that passed this test were used. Users
(or would-be users), however, reported that many of the rafts
were bursting on impact with the water. Investigation showed
that the test impact was itself weakening good rafts to the point
where they failed on a second impact. Thus, the figures collected
related to the number of rafts which had survived a particular
test in the past, not to the number which would give satisfactory
service in real emergencies.

EXAMPLE 161C   PRICE RIGIDITY

An important question of empirical research in economics is
the extent of rigidity—or flexibility—of prices in oligopolistic
industries—industries in which there are relatively few sellers.

... It is not possible to make a direct test for price rigidity, in part, because the prices at which the products of oligopolists sell are not generally known. For the purpose of such a test we need transaction prices; instead, we have quoted prices on a temporal basis and they are deficient in two respects.

The first deficiency is notorious: Nominal price quotations may be stable although the prices at which sales are taking place fluctuate often and widely. The disparity may be due to a failure to take account of quality, "extras," freight, guaranties, discounts, etc.; or the price collector may be deceived merely to strengthen morale within the industry. The various studies of steel prices . . . contain striking examples of this disparity. . . . We cannot infer that all nominally rigid prices are really flexible, but there is also very little evidence that they are really rigid.

The second deficiency is that published prices are on a temporal basis. If nine-tenths of annual sales occur at fluctuating prices within a month (as is true of some types of tobacco), and the remainder at a fixed price during the rest of the year, the nominal price rigidity for eleven months is trivial. . . .[3]

Two aspects of the quality of data are brought out by these examples. One aspect is *precision*, or reproducibility—often called, misleadingly, "reliability." In Example 161A, the numbers produced varied considerably even when there were no differences in the quantity of interest, the percentage of nitroglycerin. In Example 161B, however, the measurements were precise, in that they showed whether the rafts could withstand the test; and similarly in Example 160, the measurements were precise in that they showed exactly how many people entered the building. The second aspect is *relevance*, often called "validity." In Example 160, the numbers obtained were not relevant measures of the number viewing the exhibits at the museum, and in Example 161B the numbers were not relevant measures of performance subsequent to the test. In Example 161A, on the other hand, the numbers were perhaps relevant, in the sense of going up on the average when the true nitroglycerin content went up, and down when it went down, but they were not precise.

[3] George J. Stigler, "The Kinky Oligopoly Demand Curve and Rigid Prices," *The Journal of Political Economy*, Vol. 55 (1947), p. 69.

By the Law of Large Numbers, the average of a large number of independent measurements will be more precise and therefore have more reliability than a single measurement. On the other hand, the average will have no more relevance than the individual measurements. In the case of the box of beads described in Chap. 5, the proportion of red beads in a sample of 20 is not a very precise measure of the proportion in the box, but by averaging the results for enough samples we can get as precise a measure as we please. On the other hand, if the sampling is not random, or the beads are counted by a badly color-blind person, the average of the figures given for a large number of samples will be little if any more relevant a measure of the proportion in the box than the figure given for one sample.

There is no safety in numbers. There is something about numbers that lures people into thinking uncritically, as though the number were an intuitively obvious, crystal clear, absolutely true, inherent property of the object. This illusion ought to be dispelled; numbers should be accepted only after a careful examination of their significance. All kinds of ambiguities and complications can prevent numbers from measuring what they are supposed to measure.

In order to evaluate a set of data, it is necessary to know how they were actually obtained. Skepticism is justified when a report fails to provide these details or skims over them with such phrases as "scientific precautions were taken to insure accuracy," "by a depth interviewing technique we were able to get at true motivations, not just rationalizations," etc. Conversely, the inclusion in a report of a detailed account of the methods of measurements is a sign that the writer is at least aware of the difficulties and may therefore have done reasonably well in overcoming them. When statistical evidence is presented in popular sources, such as the daily newspapers, it would be neither appropriate nor possible, of course, to include the technical account of the data. The critical reader will then do well to ask himself the question suggested in Chapter 4, "How can they really know?"

While it is possible to become too skeptical, gullibility is far more common. It is particularly easy to be gullible when the purported evidence agrees with what is already believed, or when the facts reported admit of a plausible and ingenious interpretation which is dramatic enough to attract attention, or when the subject is remote from one's own technical knowledge. The Kinsey reports, for example, are widely quoted and discussed by scientists and laymen alike, with little or no critical study of the accuracy of their evidence.

Occasionally, information is collected which might be thought wholly inaccessible. In Chapter Three we described a study in which accurate data were obtained on the incidence of the psychoses during the last 100 years. The Federal Reserve Board has for several years been obtaining intimate details of family finances. In these examples, the methods by which the data were collected have been carefully described and checks on precision and relevance have been reported. In the absence of such documentation, however, one might well be skeptical about the possibility of obtaining such hard-to-get data.

Internal Evidence

Often data contain within themselves evidence about their own quality. Inconsistencies, irregular patterns, and unlikely or impossible values are among the common clues to poor quality, though none is ever sufficient proof of poor quality.

Inconsistencies

EXAMPLE 164   KINSEY ON MALES

A statistician reviewing a widely publicized statistical study reported:

> The number of individuals involved in the study is given on p. 10 as 12,214. On p. 5, however, there is an outline map of the United States with the legend "Sources of histories. One dot represents 50 cases." The map contains 427 dots, so presumably represents 21,350 cases. . . .
> The . . . column [of Table 41] which is said to distribute 179 males by occupation totals 237 . . . Tables 37 and 41 both include what appears to be the same distribution by age at onset of adolescence, but the frequencies differ. . . .
> Table 40, p. 198, shows . . . 11,467 . . . 30 [years] or under.

Table 41, p. 208, however, shows the number of cases from adolescence through 30 as 11,985. . . . Tables 104 and 105 . . . show fewer cases 32 years of age and under than are shown as 30 and under in Tables 40, 41, and 44. . . .

The numbers of cases shown in these clinical tables are hard to reconcile with one another, however, for the sum of the numbers shown in various subdivisions sometimes exceeds, and sometimes is exceeded by the number shown for the whole group.

. . . p. 63 . . . suggests that 300 *or less* is the number of items involved for any one person. . . . p. 50 suggests, however, that . . . the maximum history covers 521 items. . . .[4]

EXAMPLE 165 COMMUNISTS IN DEFENSE PLANTS

In this example, a general impression that the study as a whole may be highly reliable is augmented by the author's careful analysis of the degree of unreliability resulting from inconsistencies.

Suppose a respondent says he would let an admitted Communist work in a defense plant, but would fire a store clerk whose loyalty is suspected but who swears he has never been a Communist. He *could* mean both of these things. But it is a good bet that one or the other response is wrong. . . . Inconsistencies become especially serious with respect to questions about which opinion is overwhelmingly on one side. For example, only 298 people of the entire national cross-section of 4,933 said they would not fire an admitted Communist in a defense plant. But analysis shows that as many as 100 of the 298 who said they would be lenient in this situation said they would *not* be lenient in *a variety* of situations which most people think are far less dangerous. These 100 people constitute only 2 percent of our entire sample, but they constitute a *third* of all who say they would not fire a worker in a defense plant.[5]

Thus, the analysis of internal inconsistencies suggests that the proportion reported as "no" on a particular proposition was probably at least half as much again as the proportion in the sample who really meant "no."

[4] W. Allen Wallis "Statistics of the Kinsey Report," *Journal of the American Statistical Association,* Vol. 44 (1949), pp. 463–484. The paragraphing here does not conform to the original.
[5] Samuel A. Stouffer, *Communism, Conformity, and Civil Liberties: A Cross-section of the Nation Speaks Its Mind* (Garden City, New York: Doubleday and Company, Inc., 1955), pp. 46–47.

Irregularities

When averages or proportions move smoothly or regularly with respect to some related variable, at least a favorable presumption is established. The following example illustrates this kind of consistency:

EXAMPLE 166A   NEONATAL MORTALITY

The following are the death rates in 1951 of infants under one week of age:[6]

Age (days)	0–1	1–2	2–3	3–4	4–5	5–6	6–7
Deaths per 1,000 live births:	9.8	3.1	2.1	1.1	0.6	0.5	0.3

If these figures had moved erratically from day to day, instead of changing continuously in the same direction (or perhaps with one reversal of direction), it would have raised doubts about their quality.

The regularity of data by no means proves their quality, of course, for spurious regularity could be introduced in many ways. Similarly, as Example 168 shows, irregularity is not prima facie evidence of poor quality, but only of grounds for special inquiry into its causes, with inadequacy of the data prominently in mind as a possible cause. The following two examples illustrate the fruits of investigating irregularities, and the third illustrates the danger of simply assuming that irregularity proves the data are in error.

EXAMPLE 166B   ROUNDING AGES

Ages reported in the Census cluster at numbers divisible by 5, and even more at numbers divisible by 10. There is, of course, nothing in the statistics of births, deaths, and migration to confirm a true clustering at multiples of 5. Undoubtedly it reflects inaccuracies in reporting. The following table shows that clustering has diminished with time, which suggests that the data are improving in quality.

[6] *Statistical Abstract: 1954,* Table 83, p. 81.

TABLE 167

PERCENT OF POPULATION REPORTING AGES WITH EACH
FINAL DIGIT, SELECTED U. S. CENSUSES, 1880–1950.

Year	0	1	2	3	4	5	6	7	8	9	Population (Million)
1880	16.8	6.7	9.4	8.6	8.8	13.4	9.4	8.5	10.2	8.2	50
1920	12.4	8.0	10.2	9.4	9.4	11.3	9.7	9.4	10.6	9.6	106
1950	11.2	8.9	10.2	9.7	9.7	10.6	9.8	9.7	10.2	10.1	151

Source: Selected from Table 1 of an article by Ansley J. Coale, "The Population of the United States in 1950 Classified by Age, Sex, and Color — A Revision of Census Data," *Journal of the American Statistical Association,* Vol. 50 (1955), pp. 16-56.

Methods have been developed for correcting for this clustering, which, while not applied to the basic Census tabulations, undoubtedly give a more realistic picture of the ages of the population than do the basic tabulations.

EXAMPLE 167   LUMINOUS INTENSITY

The same phenomenon is often found in laboratory measurements. For example, here are measurements of the luminous intensity of 12 light bulbs:

17.70,   17.55,   17.57,   17.75,   17.50,   18.00
16.55,   16.05,   16.40,   16.40,   16.05,   16.70

All but one of these measurements terminates either in 0 or 5. This suggests that measurements were typically made only to the nearest 0.05; thus "17.55" apparently means an intensity between 17.525 and 17.575, not between 17.545 and 17.555. If this tendency to round to 0 or 5 had not been noticed, it might have appeared that the measurements were made to the nearest 0.01. It is advisable to read instruments as accurately as possible, even to the point of having to guess a little about the final digit. The information lost by unnecessary rounding cannot be restored later by statistical analysis, but the inaccuracies in the last digit do tend to average out when a number of readings are made.

EXAMPLE 168   MEASLES IN PREGNANCY

... in New South Wales at the Census of 1911 the number of deaf-mutes was curiously and conspicuously high in the age-group 10–14. This, naturally, did not escape the census statisticians at the time, but they were—not unreasonably—inclined to ascribe the excess to a more complete enumeration of the deaf at the school ages.

Ten years later, however, in the census of 1921, the peak had moved on to the age-group 20–24. In other words, it related to the same cohort. The previous explanation would not hold, and the statistician in charge was then moved to write "... there is some evidence to suggest that the increase in incidence of deaf-mutism at certain ages synchronizes with the occurrence of epidemic diseases, such as scarlet fever, diphtheria, measles, and whooping cough."

... It is desperately easy to be wise after the event, but in the census returns and in the institutional data there is clear statistical evidence which might, with the aid of an epidemiological survey, have brought to light the phenomenon of the effects of rubella [German measles] in pregnancy many years before the alert clinical mind detected it [early in the second World War, in Australia].[7]

Extreme Values

Sometimes data contain extreme values that are patently unreasonable. For example, a person's age recorded as "115" would justify suspicion. An extreme observation is a kind of internal evidence that something went wrong with the data-gathering process, at least insofar as this one observation was concerned. But under many circumstances it would not be known for sure that a given figure is impossible. An investigator who discards data that do not conform to his preconceptions is apt to end up with data tailored to those preconceptions, thus defeating the very purpose of the study. There is no point in using energy and money in collecting data, only to waste them by picking and choosing among data to satisfy preconceptions.

The tendency to discard unusual observations is too prevalent. It is better to abide by a general rule never to discard observations. Like most rules, this one has exceptions. If, while

---

[7] A. Bradford Hill, discussion, *Journal of the Royal Statistical Society, Series A (General)*, Vol. 116 (1953), p. 7.

the data are being collected, some accident or unusual event casts doubt upon a particular observation, it may be ignored. But the decision to ignore it should be made before looking at the number, and only on the grounds that the collecting process has failed; and the number should then be ignored whether or not it "looks" reasonable. (An example of good practice in this respect is provided by the vitamin experiment; see Chapter Three.) The detection of wrong observations cast doubts on the process by which the data have been produced and therefore on all other observations, even if they appear reasonable. The determination to be objective may be carried to the point of absurdity, but it seldom is; whereas all too often observations are thrown out on essentially subjective grounds.

Even when it is clearly established that certain measurements are wrong and should be excluded, it is important to be aware that discarding these observations, though justifiable, does not solve the problem of interpreting the remaining data. This is because the probability of sour measurements may be greater for certain types of observations than others. For example, a study of corporate assets would encounter the obstacle of inadequate record keeping in some small companies. Even if it were felt that information obtainable from these companies was worthless, exclusion of these companies from the sample would mean that the remaining companies would tend to be larger companies where record keeping was more satisfactory. An estimate of average assets per company would therefore tend to be too high. The statistician should at the least point out this bias and preferably should attempt to allow for it by making a special investigation of the discarded companies, or a sample of them.

EXAMPLE 169 TRIPLICATE READINGS

Suppose that three chemical analyses are made on a sample of ore, and that two of these analyses agree closely as to the concentration of a particular mineral, while the third measurement differs widely from the other two. A common laboratory procedure is to discard the apparently "wild" measurement. To illustrate the unsoundness of this procedure, we selected at random ten samples of three measurements from a population in which the mean was known to be 2. Here are the results:

Samples of Three Measurements			Mean of Two Closest Measurements	Mean of All Three Measurements
0.724	0.782	1.547	0.753	1.018[a]
1.682	1.201	0.336	1.442[a]	1.073
0.623	0.743	2.495	0.683	1.287[a]
4.334	1.663	0.045	0.854	2.014[a]
0.864	2.642	5.436	1.753[a]	2.981
2.414	1.989	2.666	2.540	2.356[a]
1.506	2.364	0.763	1.134	1.544[a]
3.048	2.037	2.759	2.904	2.615[a]
2.347	4.816	1.536	1.942[a]	2.900
2.637	2.563	1.893	2.600	2.364[a]

[a] Indicates mean closer to population mean of 2.

Five of these samples present a temptation to discard a measurement. Yet in seven of the ten samples, the mean of all three measurements is closer to the true mean of the population than is the mean of the two measurements closest to each other. This is not a sampling aberration of the particular measurements we have shown here; there are sound reasons why this kind of result is typical.[8]

EXAMPLE 170   PEARL HARBOR

An erroneous rejection of extreme observations profoundly affected the course of history when, shortly after 7 o'clock on the morning of December 7, 1941, the officer in charge of a Hawaiian radar station ignored data solely because they seemed so incredible.

[8] The samples have been formed by adding 2 to each of the first three entries from the first ten lines of the "Table of Gaussian Deviates" in The Rand Corporation, *A Million Random Digits with 100,000 Normal Deviates* (Glencoe, Illinois: Free Press, 1955). We have considered that there will be a temptation to reject the outlying observation whenever it is twice as far from the middle observation as is the remaining observation. On the average, about five-eighths of samples of three from such a population (normal) will show the greater difference more than twice the smaller difference from the middle observation. The precision obtained by averaging a certain number of values of the average-of-the-closest-two can be obtained by averaging only about half as many values of the average-of-all-three.

Spurious Regularity

The preceding three subsections have stressed irregularities. As we have seen in Chapter Five, there will be a certain amount of irregularity, that due to chance factors, even in the best statistical data. When there is less irregularity than would be expected from chance, there are grounds for suspecting that the regularity is spurious. A common cause of spurious regularity is simple dishonesty: the data have been "cooked." Those experienced in questionnaire studies are often able to detect dishonest—or grossly incompetent—interviewing by the lack of variety in recorded responses.

EXAMPLE 171   IMAGINARY COIN TOSSES

A class of 65 college students was asked to make up the outcomes of 37 imaginary tosses of a fair coin, trying to make them as realistic as possible. The results of the last five "tosses" for each student were tabulated, with the following results:

Number of Heads:	0	1	2	3	4	5	Total
Number of Students:	0	5	27	27	6	0	65

By application of the principles of probability, it can be shown that the average or expected results, if the 65 students had actually tossed fair coins, are:

Number of Heads:	0	1	2	3	4	5	Total
Number of Students:	2.0	10.2	20.3	20.3	10.2	2.0	65.0

The students thus produced too many "typical" samples—2 or 3 heads—and too few "atypical" ones—0, 1, 4, or 5 heads.

Recording Data

Care should be taken that the basic records show exactly what was observed, including any observations subsequently discarded. A sound practice, common in the natural sciences, is to record data in a way that distinguishes between numbers that were observed and those that represented the units in which these numbers are expressed. Thus, if a gas meter reads in hundreds of cubic feet and shows 2,391 on the dials, this would be reported not as 239,100 but as 2391×10^2. This indicates that the exact figure was between 239,050 and 239,150, whereas 239,100 might be interpreted as indicating an exact figure between 239,099.5 and 239,100.5. Similarly, if the last dial were judged to be a quarter of the way from 1 to 2, the record would read 2391.25×10^2, not 239,125.

When data are reported, however, they may be less accurate than the original records, as for example billions of dollars, or millions of people, if more detail would not be useful in the context of the report. In listening to election returns over the radio, for example, it is confusing to hear more than two or at most three figures—for example, 1 million, 8 hundred thousand, or perhaps 1 million, 850 thousand. Full detail, such as 1,853,428, may be so confusing, especially when there are several candidates, when several election areas are reported separately, or when several people in the room are commenting on the results as they are announced, that the hearer does not comprehend even which candidate is leading.

The numbers recorded should correspond to what is actually observed, as distinguished from what is inferred from the observed data or derived from them by computations. A few examples may help to make this point clear.

Example 172A Soap Defects

In recording the number of surface defects on bars of soap, the defects on one side were counted. On the assumption that there would be the same number on both sides, the number of defects was doubled before recording. Whether or not this assumption is correct (and it is at best true only on the average, not for each individual bar), the number of defects actually observed is the number that should be recorded. In this example, the absence of odd numbers in the data would provide internal evidence, similar to that of Example 166B (Rounding Ages), that the numbers did not correspond to the defects actually present.

Example 172B Insurance Premiums

A branch insurance office had not classified its premiums for a certain year in a way that was unexpectedly requested by the home office. The branch office assumed that the division of the total premiums into the various categories would be proportionately the same as in the previous year, and prepared figures on this basis. These data may be worse than none—for example, if the problem involved year-to-year changes in the proportions of premiums in the various categories. The central office would have been better off had it been told that the data were unavailable, and asked whether it wanted to incur the cost of getting them, or would be satisfied with estimates of the kind described; and, at least, the figures submitted should have been accompanied by a clear statement that they were estimates based on the proportions of the previous year.

EXAMPLE 173A EMPLOYMENT AND PRODUCTION

Employment is often used as a basis for estimating production. The method is a useful one; in many industries, man-hours worked provides a better index than actual finished units (recall the discussion of automobile production earlier in this chapter). But such an index of production should be accompanied by an explanation of how it was derived. Without this information there would appear to be a remarkably close relation between employment and production in these particular industries. If both the employment index and the production index are based on employment, the close correlation represents a mathematical truism rather than any significant relationship in the industrial world. A further complication in measures labeled "production" which are based on no facts except man-hours is that man-hours must be multiplied by a factor supposed to represent output per man-hour. This factor may be unreliable. Often it contains a more or less arbitrary upward trend—that is, it is increased systematically month by month (three percent per year is currently much in style). Someone comparing the output and employment figures will "discover" an upward trend in output per man-hour (called "productivity") which has simply been put there by the man who produced the output figures.

EXAMPLE 173B IMPUTED OPINIONS

A friend of the authors was interviewed in a market research study designed to find out what consumers like and do not like about toothpaste. Our friend found himself completely unable to articulate his reasons for preferring the brand of toothpaste he was using. The interviewer suggested that perhaps it was something about the taste or maybe the price, but our friend was sure these were not the reasons. Finally, when the interviewer saw that no concrete reason was forthcoming, she recorded "superior cleansing ability." The interviewer had been instructed that "don't know" was an unacceptable answer. Another market research interviewer reported that respondents were extremely inarticulate and noncommittal, but that he had been able to report specific answers because he knew intuitively what the respondents actually meant. It is entirely appropriate to give the respondent every chance to reply fully, without, of course, actually suggesting answers. If a respondent's reply is recorded (and some record should always be made, even of uninformative responses), it should be as nearly as possible as he gave it and not as "interpreted" to show what he "really" meant.

Example 174A Heart Size

To determine the size of a living person's heart from an X-ray film, a radiologist sometimes measures two diameters, multiplies them together, divides by the average ratio of this product to the projected area of the heart, multiplies by another factor to allow for parallax due to the heart's distances from the film and X-ray source, and expresses the resulting figures as a percentage of the average heart size for individuals of the given height and weight. The only direct observations in all this that have any relation to the particular patient are the two diameters. These, or possibly their product, might as well be expressed directly as a percentage of the average diameters. The intervening steps make the numbers seem more meaningful, and serve a useful purpose when data from different sources are being combined, but no amount of arithmetic will introduce any facts beyond the two diameters.

Example 174B Water Chlorination

This example, of which the consequences might have been tragic, was given us by a man who had lived for many years in the country involved:

The water supply of a certain city of about 80,000 was known to be seriously contaminated. An efficient chlorinator was, therefore, always in operation at the intake to the city system. A full-time laboratory technician was stationed in the city to measure the chlorine content of samples of water taken daily at various points throughout the city. On one occasion, however, the chlorination system failed, and various provisions against such a contingency failed also. The failure was not discovered for more than a week. During the whole period that the water supply was not chlorinated, the daily reports of the water tests continued to show approximately the same chlorine content, with only normal variation. It developed that the technician had been submitting made-up data. Fortunately, however, the waterborne-disease rate did not rise—at least the numbers purporting to measure it did not.

Conclusion

The meaningfulness of the decisions reached from a statistical analysis depends upon the relation between the numbers that are analyzed and the real world to which the decisions will relate. The numbers that are analyzed arise from a series of operations, sometimes as simple and intuitively meaningful as counting the number of times a stick can be laid along the edge of a table and writing down the number, sometimes so complicated that a vast body of scientific and technological knowledge accumulated over centuries is necessary to understand the relation between the final number and the particular aspect of the world that is of interest. In principle, the statistician, as statistician, need not be an expert on these measuring processes. In practice, his experience is likely to throw light on aspects of the measuring processes that would be overlooked by specialists without statistical training. At any rate, the user of a statistical report must inquire about the basic data before he bases any conclusions on them, and on the whole the statistician who worked with the data is best qualified to do this for him.

Even with controlled conditions, repeated measurements of the same thing typically vary somewhat. The less they vary, the more precise the measurements are said to be. However much individual measurements may vary, averages of groups of them will (by the Law of Large Numbers) vary less; the larger the groups averaged, the less the variation. If, however, the operations producing the numbers are unrelated to the subject about which information is desired, the average of a large number of observations, even though quite precise, will lack relevance. If, for example, we read a large number of clocks thinking they are thermometers, we get an average "temperature" of considerable precision but little, if any, relevance.[9]

Whether numbers are useful thus depends basically on the operations connecting them with the real world, but it is sometimes possible to get clues about their usefulness from the internal evidence within a body of data. Such clues may be provided by the self-consistency or inconsistency of different observations or averages, by the regularity or irregularity of variation in the data when they are classified in certain ways, or by the plausibility and reasonableness of the data.

For proper interpretation of data, they should be recorded exactly as observed, not as inferred from what was observed. They should be expressed so as not to confuse units of measurement with the observations: for example, if one gross (144) is the unit of measurement and 17 are observed, this should be recorded as 17 gross, not as 2,448, or if a length is measured in sixteenths of an inch it should be recorded as, say, $7\frac{5}{16}$ inches, or even 116 sixteenths, but not as $7\frac{1}{4}$ inches. Reports, of course, should show only the amount of detail pertinent, not necessarily the full detail of the original records.

[9] The wording "little, if any" is to allow for the fact that (depending on what we are measuring the temperature of) there might be some relation between the time and the temperature, in which case the clock readings would contain some information about the temperature.

Chapter 8

Kinds of Data

Univariate and Bivariate Observations

THE BASIC facts of any statistical investigation are called *observations*. Sometimes the term *items* or *scores* is used instead. The thing observed is called a *variable*. Thus, an observation is a specific value of a variable. These ideas are best brought out by illustrations:

(1) In a study of the income and spending habits of 1,000 families, the investigator obtains (among other facts) the total income for each family. Income is the variable here, and a specific family's income is a single observation; there are 1,000 such observations in all.

(2) In another study, the investigator obtains the age of the "head" of each of 1,000 families. Age of the head of the family is the variable here, and the specific ages of the heads of the specific 1,000 families are the 1,000 observations.

In each of these samples, only one variable was observed. When the observation gives only one fact (that is, refers to only one variable), it is called a univariate ("one-variable") observation. It would also be possible to obtain both incomes and ages for the same 1,000 families. In this case each observation consists of a pair of numbers and is a bivariate ("two-variable") observation. If we have 1,000 univariate observations giving ages of the head and another 1,000 univariate observations giving incomes of the same families, we cannot obtain the 1,000 bivariate observations from these. Information about how the observations are paired is required. On the other hand, a set of 1,000 bivariate observations is easily converted into two sets of 1,000 univariate observations. The terms *trivariate* and *multivariate* are obvious extensions.

The "variables" represented by observations may be of two kinds: *quantitative* and *qualitative*.

Quantitative Variables

Whenever the observations refer to a measurable magnitude, the observations are called "quantitative." Family income and age of heads of families are illustrations of quantitative variables. The numbers describing income or age have two important characteristics: (1) they have operational meanings, and (2) differences, ratios, and other results of ordinary arithmetic operations have definite meanings.

Continuous and Discrete Quantitative Variables

It is impossible to measure cash income actually received in a given year more accurately than to the nearest cent. A family can have an annual cash income of $7,457.29 or $7,457.30, but nothing in between. Since possible values of the variable, cash income, move in *discrete* steps or jumps—one cent at a time in this example—it is a *discrete* variable, and measurements of income are discrete observations.[1] Similarly, the number of leaves on a tree is a discrete variable.

In measuring age, by contrast, one is not limited in this way. Theoretically, a person's age might be measured to the nearest month, the nearest week, the nearest minute, the nearest milli-second, or any degree of fineness. Hence one may think of age as a *continuous* variable in the sense that between any pair of numbers, however close together, it is possible to have another number. Measurements, however, are limited by the precision of measuring instruments, and records are limited by the number system. It might be possible to measure and record age to the nearest minute, for example, but not much finer. Hence measurements of age actually go in discrete steps or jumps, just as do measurements of income. Regardless of whether the subject of study is described by a discrete variable, or a continuous variable, recorded measurements themselves are discrete.

Often in statistical theory the assumption that variables are continuous is made because mathematical operations are then

---

[1] An economic definition of income would call 5 cents received in a two year period an income of 2½ cents per year. When thus defined, as a *rate*, income is not a discrete but a continuous variable.

simpler—though harder for the beginner to understand. This assumption frequently gives results in good enough agreement with those obtained by more complicated methods which allow for the discreteness of actual measurements.

Quantitative Comparisons

We review here some of the arithmetical principles involved in quantitative comparisons. Joe Louis, when boxing, weighed approximately 220 pounds and Eddie Acaro, when jockeying, weighs about 110. In what ways can one compare the weights of these two men?

(1) First, Louis is 110 pounds heavier than Arcaro and Arcaro is 110 pounds lighter than Louis. This type of comparison is known as an *absolute comparison*. The "absolute difference" of weights is 110 pounds. It is possible to make absolute comparisons because weight can be expressed in equal units, pounds in this example.

(2) Louis is twice as heavy as Arcaro and Arcaro is one-half as heavy as Louis. This comparison is called a *relative comparison*. Relative comparisons require not only equal units but a definite zero point as well; for example, it is incorrect to describe a temperature of 64° F. as twice as warm as if it were 32°, for the zero point is arbitrary and does not indicate absolute lack of warmth.

Relative comparisons of this kind are frequently expressed in terms of percentages. Taking Arcaro's weight as the base of comparison, 110 pounds becomes 100 percent. Then Louis's weight of 220 pounds becomes

$$\frac{220}{110} \times 100 \text{ percent} = 200 \text{ percent.}$$

Thus, Louis's weight is 200 percent *of* Arcaro's weight, or, alternatively, 100 percent *greater than* Arcaro's weight. Both of these statements are precisely the equivalent to the earlier statement: "Louis is twice as heavy as Arcaro."

Next, make Louis's weight the base of comparison, by letting 220 pounds correspond with 100 percent. Arcaro's percentage must have the same ratio to 100 as Arcaro's weight has to Louis's weight. That is,

$$-\frac{A \text{ percent}}{100 \text{ percent}} = \frac{110 \text{ lbs.}}{220 \text{ lbs.}} \quad \text{or} \quad A \text{ percent} = 50 \text{ percent.}$$

Again the result can be expressed in two equivalent ways: Arcaro's weight is 50 percent *of* Louis's weight, or Arcaro's weight is 50 percent *less than* Louis's.

For comparisons involving percentages, five precautions are worth noting:

(1) If one quantity is 0 it is 100 percent smaller than any non-zero quantity, but one positive quantity cannot be *more* than the 100 percent smaller than another. Occasionally, such statements are encountered as: "Production of automobiles fell off 120 percent from last year." This is impossible, and it is not clear what may have been meant. Perhaps production last year was 220 percent of some base year's production, and this year returned to the base level. In other words, the base of the 120 per cent may not be last year's production.

Confusion is especially likely to occur in describing changes in percentages. If steel production rises from 80 to 90 percent of capacity, this is a relative rise of 12.5 percent or an absolute rise of 10 percent. Such confusion can be reduced by referring to the absolute rise as "10 percentage points."

It might seem reasonable to describe a change from, say, a profit of 100 in one year to a loss of 100 in the following year as a decrease of 200 percent, since the new level is below the old level by 2 times the old level. But if this definition were applied in reverse, a change from a loss of 100 to a profit of 100 would also have to be described as a decrease of 200 percent, since the new level is also obtained by subtracting 2 times the old level from the old level. It is therefore best not to use percentages to measure changes in quantities whose signs alter.

(2) Percentage changes that are equal in magnitude but opposite in sign do not offset one another. More generally, adding two successive percentage changes does not give the total percentage change. This is illustrated by the following example.

EXAMPLE 181   MOTOR VEHICLE SALES

Factory sales of motor vehicles, which were just over 8 million in 1950, declined in 1951 by nearly 1¼ million, or 15 percent. Had they then increased in 1952 by 15 percent, they would not have returned to 8 million, or 100 percent of the 1950 level, but only to 7.8 million, or 97¾ percent of the 1950 level. After the 15 percent decrease in 1951, it would have required an 18 percent increase for 1952 to regain the 1950 level. The fact is that sales decreased in 1952 by nearly 1¼ million again, but this time the decrease was not 15 but 18 percent, since the base was smaller, 6¾ instead of 8 million. And, though the 1953 figure rose more than 32 percent, this did not offset (or come within 1 percent of offsetting) the 15 and 18 percent falls of 1951 and 1952, but left sales still two-thirds of a million vehicles, or more than 8 percent, below the 1950 level.[2]

(3) Percent comparisons are awkward if one quantity is many times as large as another. The population of the United States is about 11 times as large as the population of Canada. It would be perfectly correct to say that the population of the United States is 1,000 percent larger than that of Canada, but it is hard to grasp the meaning of "1,000 percent" larger, while "11 times as large" is quite easy. When there are large relative differences between quantities being compared, percentages should be avoided.

(4) Instead of saying either that Louis's weight is 200 percent *of* Arcaro's, or 100 percent *larger than* Arcaro's, one might be tempted to state, incorrectly, that Louis's weight is 200 percent more than Arcaro's. A person reading this last statement would be led to believe that Louis is three times as heavy as Arcaro. Often misrepresentations of this kind are deliberate in propagandist writing, and once in a while they creep into disinterested studies because of ignorance or carelessness. In fact, this type of misrepresentation is so common that in the absence of the absolute numbers from which the percentages are com-

[2] The sales figures, for the benefit of those who want to calculate more precisely the changes stated in the text, were (in thousands): 1950—8003, 1951—6765, 1952—5539, 1953—7323. *Statistical Abstract: 1955*, Table 664, p. 551.

puted, there is always doubt as to whether, say, a "200 percent increase" really means a threefold or a twofold increase.

(5) The base of a percentage, when it is not explicitly stated, may not be what you take for granted. Profits expressed as percentages in the real estate business, for example, commonly represent percentages of the purchase price, but in the furniture business they represent percentages of the selling price. Thus, to buy for $1,000 and sell for $1,250 represents a 25 percent profit in real estate parlance but a 20 percent profit in furniture parlance.

Qualities or Attributes

Sometimes each observation tells only that a particular quality or *attribute*—such as "red hair," "Democrat," "over 65 years old," etc.—is present or absent. Observations of this kind are still amenable to statistical treatment, simply by counting or enumerating the number of observations in which a given quality is present and the number in which it is absent. The terms *enumeration data* and *attribute data* are sometimes applied to qualitative observations.

The results of individual psychotherapy, for example, might not be describable in terms of quantities, but the percentage of cures achieved, according to some definition of a "cure," is a number that can be interpreted statistically; so is the number of times that a single individual is cured on a particular occasion (necessarily either 0 or 1).

Sometimes qualitative observations can be ranked. The grade "A" can be ranked above the grade "B," "B" above "C," etc., for grades prepared by the same method. Attitudes might be ranked as "very favorable," "favorable," "neutral," "unfavorable," and "very unfavorable." However, there would, except for special purposes, be no meaningful way to rank the categories, "red hair," "blond hair," etc.; drug store, grocery store, clothing store, etc.; or Harvard, Yale, Columbia, Princeton, Cornell, Chicago, Stanford, etc.

Even if qualitative observations are ranked, arithmetic operations on the ranks are meaningless for describing the relations among them. For purposes of calculating "average grades" in high school or college, for example, numbers are sometimes assigned to letter grades: "F" may be —2; "D," 0; "C," 1; "B," 2; and "A," 3. Except for the fact that they should either increase or decrease from "F" to "A," these numbers are essentially misleading. It is not true, as the numbers imply, that an "A" represents three times as much of something as a "C." Someone else might prefer to attach the numbers, 0, 1, 2, 3, 4 to these grades, or 1, 2, 4, 8, 16, and there would be no basis for arguing that his scale was less valid than —2, 0, 1, 2, 3. The device may be useful as an administrative convention for designating honor students, but it should be understood that it is arbitrary, however precise and objective the arithmetic may look. With a different, equally reasonable, scoring system there would be some differences in the list of honor students.

The same general remarks apply to the field of attitude measurement. It may be useful to rank attitudes and attach numbers to them, but the resulting numbers cannot be interpreted as quantities. In particular, it is meaningless to say that Private K likes the army twice as much as Private P, or that a certain baseball fan prefers the Los Angeles Dodgers four "emotes" more than the San Francisco Giants. While it is conceivable that attitudes might be put into numbers in such a way that the units would be equal, and possibly even so that the zero point would be unique, most attempts at psychological scaling claim no more than to place results in the right order.

Obtaining Information by Communication

Much of the basic statistical data used in the social sciences, in business, and especially in public affairs is obtained by communicating with people either orally or in writing. While methods of obtaining information by communication cannot be covered adequately within the scope of this book, a few general ideas and specific rules of thumb may be pertinent. We shall consider three of the problems which must be faced: interviewing, design of the questionnaire, and summarizing individual responses.

Before going into these matters, however, it is worth while to point out that the process of getting information by communication is usually susceptible to both deliberate and unconscious abuse. The propagandist—and he is everywhere—has the choice of many subtle ways of obtaining answers which seem to prove his point. Even the objective investigator often can be misled by a mishandling of the communication process. It is not at all easy even for the expert to obtain sound measurements by communication, and the novice, who accounts for a high fraction of all questionnaires, is even more prone to error. Moreover, responses to questions, even when there are no faults of the kinds just mentioned, may be irrelevant or seriously misleading for many problems. For example, it has sometimes been concluded that businessmen do not try to maximize profits simply because they say that they do not, or even because they are not familiar with such economic concepts as "marginal cost" and "marginal revenue." Thus, in interpreting the answers to questions, as in interpreting statistical measurements generally, there is need to consider the basic issue of the relationship between numbers and the real world—in this case, the connection between what people say and what they do or what they really think.

Interviewing

The person doing the interviewing unavoidably influences the quality of the information collected. Example 96A suggested that white interviewers obtained distorted answers from Negroes on the question of the fairness of the Army toward Negro soldiers. You need only recall the times you have been interviewed to recognize how much influence the interviewer had in determining what you said. Truly skillful interviewers are scarce; not every capable person can become a good interviewer. While such qualifications as honesty and interest in the work are as essential to good interviewing as to good work in other fields, several special qualifications are required, especially two: a genuine understanding and appreciation of the need for objectivity and neutrality on the part of the interviewer, and the ability to make the respondent feel at ease so that he will be willing to make a serious attempt to answer the questions carefully. In other words, the interviewer must probe for responses without letting his own opinions intrude and without making the respondent irritated or self-conscious in a way that would lead to inaccuracies. Experience in interviewing, in the absence of proper training and of skill in putting people at ease, does not guarantee good interviewing. General intelligence and good education help, chiefly through making the interviewer aware of the need for objectivity and of the real danger that his own opinions and personality may affect the respondent's answers in subtle ways.

How can the reader of statistical reports based on surveys tell whether or not the interviewing was well done? A few rules of thumb provide some guidance. As in all aspects of statistical work, the greater the detail in which the report describes the methods used, and the fewer the vague claims, such as "highly trained interviewers," the better the work is likely to be. The description of method may tell something about the selection, training, and supervision of interviewers, and this may yield clues to the quality of the interviewing. If two or three days were devoted to training interviewers, if carefully written instructions were prepared, etc., the results are likely to be better than if the interviewers were simply given questionnaires and told to collect interviews. Finally, when one is familiar with a particular field, he is likely to learn the reputation of the organizations who do surveys in that field. The marketing research analyst usually has some idea of the interviewing methods used by firms engaged in marketing research, while the social psychologist learns the strong and weak points of the various centers for attitude and opinion research.

Questionnaires

The phrasing and sequence of questions is difficult and important. Questionnaires must always be tried ("pretested") under realistic conditions before being put into use, and often retrials are necessary to test revisions made in the light of the first trials.

(1) *Definition of Objectives.* The information desired should be defined carefully and precisely. If this essential first step is slighted, troubles will multiply as the investigation proceeds:

unnecessary or irrelevant data will accumulate, while really vital data are not obtained at all or are lost in a mountain of worksheets and questionnaires. It is necessary to confine the scope of any one questionnaire by establishing priorities for the things that are really needed over those that merely would be "interesting to find out about." Preliminary investigation, including study of previous surveys in the same field, is needed. "Problems" seldom come to the statistician in well labeled packages, while scientists frequently must devote long periods to relatively aimless exploration. But when information-gathering is formally begun, it is essential that objectives be defined clearly. Ideally, the objectives of the study should be formulated explicitly and actually enumerated in the final report.

Only after it has been decided exactly what information is wanted can methods be chosen for getting it. It is necessary to devise a series of operations by which abstract concepts are translated into actual measurements of some kind.

(2) *Ability and Willingness to Answer Questions.* Two basic requirements must be met by any question: the person who is to answer the question must be *able* to furnish the information desired and he must be *willing* to do so. Your neighbor may be willing but unable to divulge certain facts about your personal finances or your family life, while you may be able but unwilling to do so.

Unwillingness may be reflected in a high rate of "refusal" by people approached for interviews. Sometimes, also, people will permit themselves to be interviewed yet answer some questions incorrectly and refuse to answer other questions at all. The reader of a statistical report should try to find out how many people refused to be interviewed and how many did not give answers to individual questions, as evidenced by the number of responses labeled "no answer," "no response," "not ascertained," etc. Failure to report this information is a major fault in a statistical study.

Inability to furnish information is clear-cut when a person is asked about things of which he knows nothing. Special problems arise, however, when the person is asked about something of which he knows a little and can guess more, as when a husband is asked about his wife's expenditures, or when someone is asked about things nearly forgotten, such as a radio program a month ago, or experiences in childhood.

Whenever possible, objective checks should be made to find out how honestly people are replying and how able they are to give the information sought. In a market research study it was discovered that many people, asked what brand of flashlight battery they were using, named the most familiar brand when actual inspection showed they had some other.

There are many devices for overcoming inability and unwillingness. Methods of "aided recall" have been devised by psychologists to aid memory. Numerous questioning methods have been evolved to overcome unwillingness. A classic illustration of the latter is the question "When do you intend to read *Gone with the Wind?*" Actual readers replied that they had already read the book, while the rest were given an opportunity to "save face" by replying that they intended to read it soon.

(3) *Avoiding Ambiguity in Wording.* The wording of the question should be completely unambiguous—though this is sometimes nearly impossible if people of divergent backgrounds are being interviewed. Students, faced with ambiguous examination questions, to which several interpretations are logically possible, rely on their knowledge of the instructor in order to guess which answer he wants. Questionnaires are even more difficult to prepare than examinations, because respondents usually are not as strongly motivated, as well-informed about the field of the questions, or as alert and intelligent as a student taking an examination. But the goal to be sought is a wording that eliminates alternative logical interpretations and suppresses emotional overtones and unusual words which might lead to misinterpretation. It is best to use short sentences, uncomplicated sentence structure, and common, specific words. Finally, it is advantageous to explore important points by several different but related questions, since this not only gives a rounded view of the respondent's opinions but also allows for such checks of internal consistency as that described in Example 165 (Communists in Defense Plants).

EXAMPLE 188A   AIR FORCE QUESTION

In the question, "should the United States have a large air force?" the word "large" is vague. Even if the respondent were asked to choose among several sizes his answer would not have much meaning without some idea of how well he understood the costs and uses of, and the alternatives to, air weapons. To some people one thousand planes might seem a large air force; to others, 143 groups might seem small.

EXAMPLE 188B   LAW SCHOOL COURSES

Ambiguity that could easily have been avoided by pretesting the question is described in the following quotation:

> Professor Cheatham sent a questionnaire to all the law schools. In one of the questions inquiry was made as to whether the school addressed was offering a course on the professional functions of the lawyer. Approximately two-thirds of the schools answered that they were giving such a course. The remainder said they were not. He felt, however, that these figures were hardly dependable, for, from the answers given, some deemed it a "course" if the school sponsored a few lectures on the formal rules of professional conduct, while others, even if such lectures were given, did not consider that those lectures could be called a "course."[3]

EXAMPLE 188C   DESIRE FOR REFORMS

A nation-wide Gallup Poll, published August 21, 1943, presented results on the question:

> "After the war, would you like to see many changes or reforms made in the United States, or would you rather have the country remain pretty much the way it was before the war?"

> ...In view of the fact that the particular phrasing of this question seemed likely to render it especially susceptible to heterogeneous interpretations by respondents, it was decided to make a new study of this same question with the objective of ascertaining the variety of contexts in which the question tended to be answered.

> A small scale poll repeating the Gallup question was made in the New York City area between August 30 and September 4, 1943. .... In all, 114 interviews were taken.

[3] Albert J. Harno, *Legal Education in the United States* (San Francisco: Bancroft-Whitney Company, 1953), pp. 156–157.

... The procedure in each interview was as follows: The interviewer introduced himself, explaining that he was . . . making a survey among randomly chosen persons in New York. The standard poll question (see above) was then asked. Instead of ending the interview at this point, or going on to another specific question, as in customary poll practice, the interviewer then encouraged the respondent to enlarge upon his answer in an informal conversational way. The interviewer asked no direct questions at this point; he simply urged the respondent to extend and explain his answer. . . . The respondent was then asked two direct questions. If he had voted for "changes or reforms," he was asked:

a) "What sort of changes or reforms would you like to see?"

b) "Are there any changes or reforms that you wouldn't like to see?"

If he had voted for things to "remain the same," he was asked:

a) "What sort of things wouldn't you like to see changed or reformed?"

b) "Are there any changes or reforms you would like to see?"

A full verbatim account of each interview was taken, and the sex, age, income and education of the respondent recorded.

... A careful study of the verbatim reports of the free discussion following answers to the direct poll question clearly reveals a wide diversity in the interpretations which people make of the question. On the basis of the spontaneous discussion by the individual and the specific things to which he refers as requiring or not requiring changes or reforms it is possible to ascertain roughly which one of a number of frames of reference seems to be operative for each individual in responding to the poll question. Each of the 114 interviews has been rated by the authors as to frame of reference. Most of the interviews seemed to fall naturally into seven main categories; the remaining interviews were rated as non-ascertainable as to frame of reference. The seven categories . . . are as follows:

1. *Domestic changes or reforms.* This frame of reference would seem to be the one *intended* by the Gallup question. Responses classified under this heading are those referring to various social, economic and political reforms within the United States. . . .

2. *Technological changes.* Respondents classified under this heading are those who seem to be thinking of technological changes only. . . .

3. *Basic political-economic structure of the United States.* Respondents classified here are those who seem to take the question to refer to drastic changes in the United States, such

as in the Constitution, in democracy, in "our way of life"....

4. *Foreign affairs of the United States.* These respondents seem to interpret the question as referring to changes in our relations with other countries....

5. *Immediate war conditions as the standard of judgment.* Some respondents seem to answer the question in terms of a comparison of the peaceful state of affairs before the war with the current unpleasant war conditions. They do not really express an opinion about *post-war* changes or reforms....

6. *Immediate personal condition as the standard of judgment.* In these interviews the answers are made narrowly in terms of a consideration of whether the individual himself is satisfied, not of whether changes or reforms in general in the country are desirable....

7. *Desirable state of affairs in general as the standard of judgment.* In these cases the question is parried. No real content is embodied in the answer; there is merely an affirmation of the desirability of a good state of affairs....

Frame of reference non-ascertainable....

The percentage of people giving each interpretation to the poll question (and hence the relative importance of each frame of reference in determining the final poll tabulation) is given [below]. It appears that roughly one-third of the sample of respondents interprets in frames of reference other than "Domestic changes or reforms," which is, presumably, the *intended* frame of reference for the question.

Frame of Reference	Percentage of respondents
Domestic changes or reforms	63%
Technological changes	2
Basic political-economic structure of U. S.	10
Foreign affairs of the U. S.	3
Immediate war conditions as standard of judgment	4
Immediate personal condition as standard of judgment	7
Desirable state of affairs in general as standard of judgment	4
Non-ascertainable	7
	100%
	$N = 114$

The analysis proceeds to show, among other things, that the desire for "changes or reforms" depended on what kind of changes or reforms the respondent had in mind.[4]

[4] Richard S. Crutchfield and Donald A. Gordon, "Variations in Respondents' Interpretations of an Opinion-Poll Question," *International Journal of Opinion and Attitude Research*, Vol. 1 (1947), pp. 1–12.

(4) *Neutrality of Wording.* Answers will vary with the wording of questions even though the literal meaning is the same. Consider the following two questions:

"You approve of rent control, don't you?"

"You don't approve of rent control, do you?"

Probably a larger proportion of people would indicate approval of rent control in reply to the first question than to the second, yet literally interpreted the two questions ask the same thing. A more "neutral" wording, such as "Do you approve or disapprove of rent control?" would be preferable. However, no wording is correct in any ultimate sense: the purpose of the study must be considered. If a study purporting to show opposition to rent control had employed the question, "You don't approve of rent control, do you?" its conclusions would be, to say the least, questionable. It is possible, however, that for certain purposes this question might be quite legitimate—for example, to find out how many people are strongly enough in favor of rent control to say so even when asked a leading question suggesting the opposite.

Sometimes surveys use different forms of a question with different respondents. If the results are fairly similar for the different forms, this is often interpreted as showing that opinion is fairly well crystallized, hence not subject to slight influences. Low proportions not answering or answering "don't know" (referred to as NA's and DK's in the jargon of the opinion surveyors) also are sometimes interpreted this way.

The sequence of questions, as trial lawyers know, may affect responses. If a respondent has just answered that he thinks his landlord would increase his rent if rent controls were removed, he may be more inclined to say that he favors rent control.

(5) *Rule-of-Thumb Checks.* Before studying the results of a survey, the wording and sequence of questions of the questionnaire should be examined. While specialized experience with questionnaires is needed for real expertness, common sense alone is often good enough to detect serious errors.

One good rule-of-thumb is to answer the questions yourself, trying to put yourself in the respondent's place if the questions do not apply to you, to see if you would understand them completely and would be able and willing to answer them. If the questionnaire fails this test, there is reason to question the information obtained from it. If it passes this test, it still may have deficiencies as an instrument for obtaining information from a "cross-section" of the population. College graduates, for example, would probably have no trouble in understanding the expression "frame of reference," yet it would be meaningless to many people. So if you understand the questions yourself and would be willing and able to answer them, you must then ask if the information requested is of the kind that most people would be willing and able to give.

Coding

Questions may be grouped into two categories with respect to the way in which they are to be answered: "fixed-alternative," and "open-ended" or "free-response." In the fixed-alternative question, the respondent is given a limited number of responses, such as "yes," "no," and "don't know." In the free-response question, the respondent answers in his own words and the interviewer records them as nearly verbatim as he can. Sometimes the two methods are combined; a person might be asked if he approved or disapproved of rent control (fixed-alternative), and then asked for his reasons (free-response).

Regardless of the type of question used, there is a problem in classifying the respondent's answer. With the fixed-alternative question, this classification is done jointly by the question-writer, who establishes the categories, and the interviewer, who puts the responses into the categories. With the free-response question the answers actually recorded by the interviewer must be classified later into a relatively small number of groups so that the main findings can be summarized. The technical name for the process of classifying answers to free-response questions is "coding." Some understanding of this process is useful in interpreting statistical findings. A few standard requirements of coding may be mentioned.

(1) Each answer should fit into at least one category. In other words, the categories ought to be exhaustive.

(2) Each answer should fit into only one category. In other

words, the categories ought to be mutually exclusive. This condition, while desirable, is not essential and is often not feasible. If, for example, the response to "What kind of things do you worry about most?" is "Paying bills. My husband has been in the hospital and may have to go back again," it might be classified either as "Personal Business or Family Finance" or as "Health of Self and Family" if these are two of the categories.[5] No set of categories is likely to avoid such double classifications for an "open-ended" question of this kind. The drawback to having a single answer classified in more than one category is that the percentages, in the tabulation of number of respondents giving each answer, will exceed 100, thus complicating comparisons among questions which differ in the total number of responses given, and tending to overemphasize the views of respondents who give longer or more comprehensive answers, which are classified under several headings.

(3) If the question itself is ambiguous, such as "Do you think you are better or worse off than before the war?" the separate interpretations of the question (personal health and happiness, national affairs, etc.) should be kept separate. The answers of those who thought the question referred to national affairs, for example, should if possible be tabulated separately from the answers of those who thought it referred to personal affairs. This will bring into the open the fact that the question was ambiguous, and make it possible to salvage something. The answers reveal, at least, the things people think of when "well-being" is mentioned.

(4) The categories should be chosen for their pertinence to the subject being studied. Ordinarily responses would not be classified according to their length, for example—though they are so classified in some psychological tests, for example, the Rorschach test.

(5) The number of categories used must represent a compromise between the need for summarization and the need for

---

[5] The question and response are taken from an actual survey made in May, June, and July, 1954. The answer was actually classified as concern over personal business or family economic problems. Samuel A. Stouffer, *Communism, Conformity, and Civil Liberties: A Cross-section of the Nation Speaks Its Mind* (Garden City, New York: Doubleday and Company, Inc., 1955), p. 61.

knowing the nuances and fine shades of meaning in the individual responses.

(6) The process of coding should be "reliable" in the sense that different people would agree pretty well on the category into which each response should be classified.

(7) The number of people who did not answer the questions or whose answers were not recorded by the interviewer should be shown in a "no answer" or "not ascertained" category. Even if this category is not shown in the report of a study, there usually are some responses of this kind. It is often possible to infer from the internal evidence of the findings whether or not there were "no answers." If there is a large number of "no answers," allowance must be made for the fact that people who do not answer are frequently different from those who do. In the 1948 presidential pre-election surveys, for example, a large number of people were undecided, and there is evidence that these people voted more heavily for Truman than did the group of people who had made up their minds.

Conclusion

Observations are classified as univariate, bivariate, trivariate, or multivariate, according to the number of variables observed jointly. They are classified as quantitative or qualitative, according as they record measurements (or counts) or simply nonquantitative attributes. Quantitative variables are classified as continuous or discrete, according to whether any value is possible within a certain range, or only particular values; actual recorded measurements are always discrete, even when the variable measured is continuous. Qualitative variables are dealt with statistically by counts of the number of times they occur.

A kind of measurement of general public interest, about which statisticians are especially expected to be informed, is that occurring when information is obtained by asking people questions. Such surveys require expert interviewing; careful attention to the questions, to bring out the information really wanted, to ask about things that people will be willing and able to answer, to avoid ambiguity and to avoid influencing answers; and careful coding of answers for statistical summarization.

Chapter 9

How to Read a Table

INFORMATION CAN be packed into a statistical table like sardines into a can, and if you cannot read a table, it is as if you had a can of sardines but no key. Ordinary reading ability is no more effective in reading a table than an ordinary can opener in opening a can of sardines, and if you go at it with a hammer and chisel you are likely to mutilate the contents.

We will try to extract information from Table 196 about the association of illiteracy with age, color, and sex. We urge that before you read further you study Table 196 and jot down your own conclusions in the sequence in which you reach them.

EXAMPLE 195   ILLITERACY

You will not extract any information from the table if you continue to divert your gaze from it in embarrassed bewilderment. Don't stare at it blankly, either—focus your eyes and pick out some detail that is meaningful, then another, then compare them, then look for similar comparisons, and soon you'll know what the table says.

There are at least two good reasons for learning to read tables. The first is that once the reading of tables is mastered (and this does not take long), the reader's time is greatly economized by reversing the usual procedure, that is, by studying the tables carefully and then just skimming the text to see if there is anything there that is not evident in the tables, or not in them at all. This not only saves time but often results in a better understanding: a verbal description of any but the simplest statistical relationship is usually hard to follow, and besides, authors sometimes misrepresent or overlook important facts in their own tables. A second reason for learning to read tables is that users of research can better describe the data needed to answer their administrative or scientific problems if they can

TABLE 196

Illiteracy Rates, by Age, Color, and Sex, 1952

Based on a sample of about 25,000. Persons unable both to read and to write in any language were classified as illiterate, except that literacy was assumed for all who had completed 6 or more years of school. Only the civilian, noninstitutional population 14 years of age and over is included.

Percent Illiterate

Age (years)	White			Nonwhite			Both Colors		
	Male	Female	Both	Male	Female	Both	Male	Female	Both
14 to 24	1.2	0.5	0.8	7.2	1.4	3.9	1.8	0.6	1.2
25 to 34	0.8	0.6	0.7	9.7	3.8	6.4	1.6	0.9	1.2
35 to 44	1.2	0.5	0.8	7.5	5.9	6.6	1.7	1.0	1.3
45 to 54	2.2	1.4	1.8	12.8	10.4	11.5	3.2	2.3	2.7
55 to 64	3.6	3.4	3.5	19.4	16.9	18.1	4.7	4.4	4.5
65 and over	5.6	4.4	5.0	35.8	31.2	33.3	7.6	6.2	6.9
14 and over	2.1	1.5	1.8	12.7	8.2	10.2	3.0	2.1	2.5

Source: Statistical Abstract: 1955, Table 132, p. 115. Original source: Bureau of the Census, *Current Population Reports*, Series P-20, No. 45.

specify the types of tables needed, and this requires an understanding of tables. Research workers, in turn, can plan investigations more effectively if they visualize in advance the statistical tables needed to answer the general questions that motivate the research.

Consider, then, Table 196. By following a systematic procedure it is possible to grasp quickly the information presented. Here are the main steps:

(1) *Read the title carefully.* One of the most common mistakes in reading tables is to try to gather from a hit or miss perusal of the body of the table what the table is really about. A good title tells precisely what the table contains. In this case, the title shows that the table tells about illiteracy, in relation to age, color, and sex, in 1952, and that the data are presented as rates—percent illiterate.

(2) *Read the headnote or other explanation carefully.* In the headnote to Table 196 we get a more precise indication of the basis for classifying people as illiterate. We see, in fact, that the rates are slightly too low because it was taken for granted that any person who had completed six or more years of school was literate; but it is reasonable to suppose that the error from this source is negligible. We note also that the mentally deficient, criminals, and others in institutions have been excluded, as have the armed forces, so that the data relate to people in everyday civilian life. Finally, we note that the data are based on a sample, so we make a mental note not to attach too much importance to any single figure, or difference between figures, without first looking up the sampling error.

Information of the kind given in the headnote of Table 196 is often not attached directly to the table, but must be sought elsewhere in the text. Those who prepare reports that include statistical tables should, but frequently do not, keep in mind not only the reader who reads straight through the report without putting it down, but also the user making a quick search for a specific piece of information.

(3) *Notice the source.* Is the original source likely to be reliable? In this case, the answer is definitely "yes," for the Bureau of the Census is one of the most competent statistical agencies

in the world. The secondary source, the *Statistical Abstract*, is a model of its kind. But *you* are getting the data from a tertiary source, this book. What about its reliability? Unless you have checked some of our previous data against their sources, you really do not know about that, and even if you did it would be a mistake to put complete reliance on the data without verifying them.[1] Of course *we* assure you of our reliability; but we would not trust your infallibility, or even our own, no matter who gave *us* assurances.

(4) *Look at the footnotes.* Maybe some of them affect the data you will study. Sometimes a footnote applies to every figure in a row, column, or section, but not every figure to which it applies has a footnote symbol.

(5) *Find out what units are used.* Reading thousands as millions or as units is not uncommon. Long tons can be confused with short tons or metric tons, meters with yards, degrees with radians (as in Example 112). U. S. with Imperial gallons, nautical with statute miles, rates per 1,000 with rates per 100,-000, "4-inch boards" with boards 4 inches wide,[2] fluid ounces with ounces avoirdupois, and so on. In Table 196 illiteracy is expressed in percent—incidence per 100—and age in years.

The foregoing steps are, in a sense, all for preliminary orientation before settling down to our real purpose—as a dog turns around two or three times before settling down for a nap. They do not take long, and ought to be habitual, but if you omit them you may suffer a rude awakening later—or never awaken at all.

(6) *Look at the over-all average.* The illiteracy rate for all ages, both colors, and both sexes—the whole population, in other words—is shown in the lower right hand corner of Table 196 as 2.5 percent, or one person in 40. This may surprise you, for probably not one in 400 and perhaps not even one in 4,000 of your acquaintances 14 years of age or older is illiterate. On

---

[1] Please let us know of the inaccuracies you find, here or elsewhere in this book.
[2] A "4-inch board" is 3¾ inches wide, the 4 inches referring to the width of the rough lumber.

A useful compilation of units in common use is *World Weights and Measures: Handbook for Statisticians*, prepared by the Statistical Office of the United Nations in collaboration with the Food and Agriculture Organization of the United Nations (provisional ed.; New York: United Nations, 1955).

a matter like this, for a country of 185 million people and four million square miles, neither one's own impressions nor the consensus of one's friends' impressions is valid.

(7) *See what variability there is.* It is quickly evident that there are percentages less than 1 and more than 30 in the table. There is, therefore, extraordinary variation in illiteracy among the 24 basic groups into which the population has been divided (two sexes, two colors, six age classes).

(8) *See how the average is associated with each of the main criteria of classification.*

(a) *Age.* Looking in the section for "both colors" and down the column for "both" sexes, we see that the illiteracy rate is essentially constant at about 1¼ percent from ages 14 to 44, but then rises sharply through the remainder of the age classes to a rate in the highest age class 5.7 percentage points larger than, and 5¾ times as large as, the rate in the lowest age class. (Avoid phrases such as "illiteracy increases with age," which suggest that given individuals change as they age.)

At this point, some competent table-readers, especially if they were particularly interested in the association between age and illiteracy, would pursue this path further. We shall, however, complete our survey of the gross associations with the three variables, then take up each in detail. Probably neither route has any general advantage over the other.

(b) *Sex.* In the "both colors" section, comparison of the entries at the bottoms of the "male" and "female" columns, which apply to all ages, shows that the illiteracy rate for males (3.0 percent) is over 40 percent larger than that for females (2.1 percent). In view of our finding about age, we make a mental note to consider the possibility that is is merely the association with age showing up again in the guise of a sex difference, through the medium of a difference in the age distributions of the sexes. Correspondingly, we make a note to check on the possibility that the apparent association with age is due to differences in the sex ratio at different ages. More generally, we recognize that the associations with age and sex may be *confounded,* that is, mixed together in what looks like an association with age and an association with sex.

The idea of confounding is important enough for a digression. Suppose illiteracy rates by sex and age were:

Age	Male	Female	Both Sexes
Young	1.0	1.0	1.0
Old	10.0	10.0	10.0

These hypothetical illiteracy rates are identical for young males and young females. They are also identical for old males and old females. But they differ greatly between the young and the old. In other words there is a strong relation between age and illiteracy, but none at all between sex and illiteracy. Now suppose that a group of 700 people is divided by age and sex as shown below:

Age	Male	Female
Young	100	300
Old	200	100

The over-all illiteracy rate for males would be

$$\tfrac{1}{3} \times 1.0 + \tfrac{2}{3} \times 10.0 = 7.0;$$

for females it would be

$$\tfrac{3}{4} \times 1.0 + \tfrac{1}{4} \times 10.0 = 3.25.$$

Males show a higher over-all illiteracy rate, simply because relatively more of the males are old (two-thirds instead of one-fourth) and the illiteracy rate is higher for the old of either sex. In such a case, the age and sex effects are said to be *confounded*. That is, what is really an age effect appears in the totals as a sex effect, because the age effect has had a different influence on the two sexes due to their different age distributions.

It is usual in statistics to refer to an association with, say, age, as an "age effect," or as the "effect of age," without intending the cause-and-effect implication that this term tends to carry in ordinary usage. All that is meant in statistics is association, and we will use the term "effect" that way.

(c) *Color.* To see the effect of color, we compare the entries at the bottoms of the "both" sexes columns in the "white" and "nonwhite" sections, and find the nonwhite rate (10.2 percent) to be 5⅔ times the white rate (1.8 percent). Again, however, we resolve to investigate possible confounding of all three effects.

The main effects, then, seem to be that *illiteracy rates are higher for older people, for males, and for nonwhites.*

(9) *Examine the consistency of the over-all effects and the interactions among them.*

(a) *Age.* The increase of illiteracy with age holds separately for whites and nonwhites. Some difference in detail does appear. For one thing, the nonwhite rate is not constant from ages 14 to 34, but is noticeably lower from 14 to 24. More conspicuous, the increase from the lowest to the highest age class is much larger for nonwhites than for whites: the differences are 29.4 percent and 4.2 percent, and the ratios[3] 8.5 and 6.2. Thus, it appears that age has a greater effect on illiteracy for nonwhites than for whites. For the two sexes, on the other hand, age has about the same effect, as measured by the absolute change (5.8 percent for males and 5.6 percent for females) from the lowest to the highest age class; since females have a lower rate, this makes the ratio higher for females (10.3) than for males (4.2).

A still more careful study of the table would test whether these conclusions hold if we compare, say, the next-to-lowest age class with the next-to-highest (the conclusions are the same), thus guarding against aberrations in individual cases.

Before we italicize these conclusions derived from comparing the separate section totals, let us see whether they hold within sections, that is for each sex of a color, or for each color of a sex. Here, for the first time, we use the real core of the table, the rates for the 24 basic cells. Heretofore we have used only data combined by age, by sex, or by color, or by two of these, or (in step 6) by all three.

[3] Ratios are not very satisfactory for describing changes in percentages unless the percentages remain small, because of the fixed upper bound of 100. The nonwhite rate of 33.3 percent at 65 years and over, for example, could not be multiplied by 8.5 again. Furthermore, the ratios depend on which percentage is used, that for occurrences or that for nonoccurrences. The literacy rates corresponding with the illiteracy rates mentioned in the text, while they have the same numerical differences as the illiteracy rates, have the ratios 1.44 and 1.04.

First, compare the males of the two colors. Then compare the females. Both comparisons confirm the conclusion that *the increases in illiteracy associated with increases in age are greater for nonwhites than for whites* and that *they are about the same for males as for females.* These statements are equivalent to saying that *the excess of nonwhite over white illiteracy rates is greater in the older age classes* and that *the difference between the sexes is not systematically related to age.*

(b) *Sex.* Similar detailed study leads to the conclusion that *the excess of the male over the female rate is higher for nonwhites than for whites.* Put the other way around, this says that *the difference between the colors is larger for males than for females.*

(c) *Color.* Our conclusions about the interaction between color and sex and between color and age have already been recorded in discussing age and sex.

10) *Finally, look for things you weren't looking for—aberrations, anomalies, or irregularities.* The most interesting irregularity that we have noticed in Table 196 is in the age class 25–34. For white males this is below—in fact, one-third below—the rates for the preceding and following age classes. For the nonwhite males, however, the rate is above that of the adjacent age classes by about one-third. (The white females also show a higher rate in this age class than in the adjacent ones, but only by 0.1, which might be almost all due to rounding the figures to the nearest tenth of a percent, and in any case is less than the necessary allowance for sampling error.) In attempting to form a plausible conjecture to explain this peculiarity, we first note that the period when this age class was at ages 6 to 8, and therefore learning to read and write, was 1924 to 1935. This suggests nothing to us, though it might to an expert on the subject matter. As a second stab, we note that during the period of World War II, 1942–45, this age class was 15 to 27 years old. It is, therefore, the group that provided the bulk of the armed forces. This lead seems worth investigating. Did the armed forces teach many illiterates to read and write? If so, did this affect white males more than nonwhite? Even so, why would the rate for nonwhite males be increased? Could it be that

mortality among whites was higher for illiterates than for literates, but for nonwhites the reverse? We should be surprised if any of these is the explanation, but investigating them would probably lead us to the explanation. A possible explanation, of course, is that the aberration is due to sampling error, or even clerical or printing error, and that the search for substantive explanations would be in vain. But such anomalies are often worth pursuing; this is one of the secrets of serendipity, from which the most fruitful findings of research often result. We would certainly pursue these questions if we were investigating illiteracy instead of explaining how to read a table.

In summary, then, here is what can be read from Table 196, and in considerably less time than it has taken us to tell about it:

Illiteracy in 1952 among the civilian, noninstitutional population 14 years of age and older—

(i) Averaged 2.5 percent.

(ii) Varied greatly with age, color, and sex.

(iii) Was higher at the higher ages, for nonwhites, and for males, with

    (a) the age differences higher for nonwhites—that is, the color differences larger at the higher ages;

    (b) the sex difference larger for nonwhites—that is, the color differences larger for males;

    (c) no interaction between age and sex.

(iv) Was, in the 25–34 year age class, anomalously lower for white males, but higher for colored males, than in the age classes just above and just below.

EXAMPLE 204   BRAINS AND BEAUTY AT BERKELEY

Repeating the same steps in reading Table 204 as in reading Table 196, we find at stage 8 that grades are higher in later years in college and with poorer appearance (which, to repeat earlier warnings, does not necessarily mean that given coeds get better grades as they progress in college or regress in appearance). At stage 9, however, we find it necessary to introduce such strong qualifications to the appearance effect as almost to withdraw the finding. All

TABLE 204

Mean Grades of College Women, by Appearance and Year in College

Data on 643 women students of the University of California who had completed two or more years of college, classified by beauty of face. Grades averaged by scoring A as 3, B as 2, C as 1, D as 0, E or F as — 1. Frequencies on which averages are based are shown in Table 205.

Year	Homely	Plain	Good Looking	Beautiful	All Appearances
Junior	1.58	1.45	1.34	1.16	1.37
Senior	1.56	1.52	1.45	1.57	1.50
Graduate	1.67	1.70	1.70	1.53	1.68
All Years	1.62	1.56	1.44	1.42	1.51

Source: S. J. Holmes and C. E. Hatch, "Personal Appearance as Related to Scholastic Records and Marriage Selection in College Women," *Human Biology*, Vol. 10 (1938), pp. 65–76. The means shown here have been recomputed from the original data, loaned by the authors, and in a few instances differ by one unit in the last decimal place from those given in the source.

we can say is that for juniors grades decrease with better appearance, but for seniors and graduate students there is no systematic relation. The main effect of appearance is partly a manifestation of the year-in-college effect, in conjunction with different distributions by appearance for the three college classes.

The mean for the plain, for example, is

$$\frac{68}{250} \times 1.45 + \frac{100}{250} \times 1.52 + \frac{82}{250} \times 1.70 = 1.56$$

and for the good looking it is

$$\frac{108}{236} \times 1.34 + \frac{84}{236} \times 1.45 + \frac{44}{236} \times 1.70 = 1.45$$

where the frequencies are from Table 205.[4] The difference between these two means is partly due to the fact that the juniors, who have the lowest grades in both appearance groups,

[4] The discrepancy between the result just obtained, 1.45, and the corresponding number in Table 204, 1.44, is due to the fact that the row and column averages in Table 204 were computed from data more accurate than those shown in the body of the table.

TABLE 205

COLLEGE WOMEN, BY APPEARANCE AND YEAR IN COLLEGE

See headnote to Table 204.

Year	Homely	Plain	Good Looking	Beautiful	All Appearances
Junior	17	68	108	25	218
Senior	27	100	84	33	244
Graduate	39	82	44	16	181
All Years	83	250	236	74	643

Source: Same as Table 204.

constitute 46 percent of the good looking and only 27 percent of the plain. Similarly, the graduates, who have the highest scores in both appearance groups, constitute 33 percent of the plain but only 19 percent of the good looking. Thus, the difference between these two appearance groups is partly due to the fact that the class effect operates differently in one than the other. The difference between the averages for the plain and good looking is not wholly due to the class effect, however, for among the plain the average for each class is as high as or higher than the average for the same class among the good looking.

Since the appearance effect is not present for the seniors or graduates, we conclude that its presence for all classes combined reflects partly the effect for the juniors and partly confounding of the class effect—that is, heavier representation in some appearance groups than in others of those classes which receive low grades. It would be possible for the appearance effect to work in one direction in all three classes, but in the opposite direction for all classes combined.

EXAMPLE 205   CAR PURCHASE PLANS

In reading Table 206 there are two things to be kept in mind besides things of the kinds already mentioned. (*i*) The income classes are so broad that even within a class average income

TABLE 206

PERCENT OF CONSUMER UNITS PLANNING IN 1948 TO
PURCHASE A NEW CAR IN 1949 AND PERCENT PUR-
CHASING ONE, BY EDUCATION AND 1948 INCOME

A consumer unit is defined as all persons living in the same dwell-
ing, and belonging to the same family, who pool their income to
meet their major expenses. Data are for nonfarm units and are from
a sample survey covering about 2,500 farm and nonfarm units, conducted by the Federal Reserve System in cooperation with the University of Michigan Survey Research Center.

1948 Income	Education of Head of Unit		
	Grammar School	High School	College
Under $2,000			
Planned	1	2	4
Purchased	1	4	8
$2,000 to $4,999			
Planned	4	7	15
Purchased	5	8	11
$5,000 and over			
Planned	20	21	26
Purchased	16	24	27

Source: Irving Schweiger, "The Contribution of Consumer Anticipations in
Forecasting Consumer Demand," in National Bureau of Economic Research,
Short-Term Economic Forecasting, Studies in Income and Wealth, Vol. 17
(Princeton: Princeton University Press, 1955), p. 461.

may, and undoubtedly does, increase substantially with education; hence the education effect is not completely separated from
the income effect. (*ii*) Of those with, say, under $2,000 in-
come, some were in this income class in 1948 only because of
temporarily low incomes. Such consumers probably base purchase plans, and also purchases, to a considerable degree on
expected or "normal" income. Hence if "normal" income could
have been used in place of actual income in 1948, the variation
among income classes (and, correspondingly, among education

groups within broad income classes) would have been more than is shown in the table.

Conclusion

How to read a statistical table perceptively can be learned easily by adopting systematic procedures. The first steps are orienting ones, finding out from the title, headnote, footnotes, source notes, and other explanatory matter, as well as from the row and column headings, the nature of the information tabulated. A second group of steps involves starting with simple, over-all features of the table: the general level of the entries, their variability, and the association indicated by the summary, or marginal, rows and columns. Next, the core of the table is examined to see if the effects suggested by the marginal rows and columns are true within separate rows and columns, and if so whether to the same extent. Finally, the table is examined for any unusual relations or entries that may suggest ideas or questions other than those with which the table was approached.

INDEX

Index